SpringerBriefs in Computer Science

T0205813

For further volumes:
http://www.springer.com/series/10028

Ágnes Vathy-Fogarassy
János Abonyi

Graph-Based Clustering and Data Visualization Algorithms

 Springer

Ágnes Vathy-Fogarassy
Computer Science and Systems Technology
University of Pannonia
Veszprém
Hungary

János Abonyi
Department of Process Engineering
University of Pannonia
Veszprém
Hungary

ISSN 2191-5768 ISSN 2191-5776 (electronic)
ISBN 978-1-4471-5157-9 ISBN 978-1-4471-5158-6 (eBook)
DOI 10.1007/978-1-4471-5158-6
Springer London Heidelberg New York Dordrecht

Library of Congress Control Number: 2013935484

Printed on acid-free paper

Springer is part of Springer Science+Business Media (www.springer.com)

Preface

Clustering, as a special area of data mining, is one of the most commonly used methods for discovering hidden structure of data. Clustering algorithms group a set of objects in such a way that objects in the same cluster are more similar to each other than to those in other clusters. Cluster analysis can be used to quantize data, extract cluster prototypes for the compact representation of the data set, select relevant features, segment data into homogeneous subsets, and to initialize regression and classification models.

Graph-based clustering algorithms are powerful in giving results close to the human intuition [1]. The common characteristic of graph-based clustering methods developed in recent years is that they build a graph on the set of data and then use the constructed graph during the clustering process [2–9]. In graph-based clustering methods objects are considered as vertices of a graph, while edges between them are treated differently by the various approaches. In the simplest case, the graph is a complete graph, where all vertices are connected to each other, and the edges are labeled according to the degree of the similarity of the objects. Consequently, in this case the graph is a weighted complete graph.

In case of large data sets the computation of the complete weighted graph requires too much time and storage space. To reduce complexity many algorithms work only with sparse matrices and do not utilize the complete graph. Sparse similarity matrices contain information only about a small subset of the edges, mostly those corresponding to higher similarity values. These sparse matrices encode the most relevant similarity values and graphs based on these matrices visualize these similarities in a graphical way.

Another way to reduce the time and space complexity is the application of a vector quantization (VQ) method (e.g. k-means [10], neural gas (NG) [11], Self-Organizing Map (SOM) [12]). The main goal of the VQ is to represent the entire set of objects by a set of representatives (codebook vectors), whose cardinality is much lower the cardinality of the original data set. If a VQ method is used to reduce the time and space complexity, and the clustering method is based on graph-theory, vertices of the graph represent the codebook vectors and the edges denote the connectivity between them.

Weights assigned to the edges express similarity of pairs of objects. In this book we will show that similarity can be calculated based on distances or based on structural information. Structural information about the edges expresses the degree of the connectivity of the vertices (e.g. number of common neighbors).

The key idea of graph-based clustering is extremely simple: compute a graph of the original objects or their codebook vectors, then delete edges according to some criteria. This procedure results in an unconnected graph where each subgraph represents a cluster. Finding edges whose elimination leads to good clustering is a challenging problem. In this book a new approach will be proposed to eliminate these inconsistent edges.

Clustering algorithms in many cases are confronted with manifolds, where low-dimensional data structure is embedded in a high-dimensional vector space. In these cases classical distance measures are not applicable. To solve this problem it is necessary to draw a network of the objects to represent the manifold and compute distances along the established graph. Similarity measure computed in such a way (graph distance, curvilinear or geodesic distance [13]) approximates the distances along the manifold. Graph-based distances are calculated as the shortest path along the graph for each pair of points. As a result, computed distance depends on the curvature of the manifold, thus it takes the intrinsic geometrical structure of the data into account. In this book we propose a novel graph-based clustering algorithm to cluster and visualize data sets containing nonlinearly embedded manifolds.

Visualization of complex data in a low-dimensional vector space plays an important role in knowledge discovery. We present a data visualization technique that combines graph-based topology representation and dimensionality reduction methods to visualize the intrinsic data structure in a low-dimensional vector space.

Application of graphs in clustering and visualization has several advantages. Edges characterize relations, weights represent similarities or distances. A Graph of important edges gives compact representation of the whole complex data set. In this book we present clustering and visualization methods that are able to utilize information hidden in these graphs based on the synergistic combination of classical tools of clustering, graph-theory, neural networks, data visualization, dimensionality reduction, fuzzy methods, and topology learning.

The understanding of the proposed algorithms is supported by

- figures (over 110);
- references (170) which give a good overview of the current state of clustering, vector quantizing and visualization methods, and suggest further reading material for students and researchers interested in the details of the discussed algorithms;
- algorithms (17) which aim to understand the methods in detail and help to implement them;
- examples (over 30);
- software packages which incorporate the introduced algorithms. These Matlab files are downloadable from the website of the author (www.abonyilab.com).

The structure of the book is as follows. Chapter 1 presents vector quantization methods including their graph-based variants. Chapter 2 deals with clustering. In the first part of the chapter advantages and disadvantages of minimal spanning tree-based clustering are discussed. We present a cutting criteria for eliminating inconsistent edges and a novel clustering algorithm based on minimal spanning trees and Gath-Geva clustering. The second part of the chapter presents a novel similarity measure to improve the classical Jarvis-Patrick clustering algorithm. Chapter 3 gives an overview of distance-, neighborhood- and topology-based dimensionality reduction methods and presents new graph-based visualization algorithms.

Graphs are among the most ubiquitous models of both natural and human-made structures. They can be used to model complex structures and dynamics. Although in this book the proposed techniques are developed to explore the hidden structure of high-dimensional data they can be directly applied to solve practical problems represented by graphs. Currently, we are examining how these techniques can support risk management. Readers interested in current applications and recent versions of our graph analysis programs should visit our website: www.abonyilab. com.

This research has been supported by the European Union and the Hungarian Republic through the projects TMOP-4.2.2.C-11/1/KONV-2012-0004—National Research Center for Development and Market Introduction of Advanced Information and Communication Technologies and GOP-1.1.1-11-2011-0045.

Veszprém, Hungary, January 2013   Ágnes Vathy-Fogarassy
                                                                     János Abonyi

References

1. Jaromczyk, J.W., Toussaint, G.T.: Relative neighborhood graphs and their relatives. Proc. IEEE **80**(9), 1502–1517 (1992)
2. Anand, R., Reddy, C.K.: Graph-based clustering with constraints. PAKDD 2011, Part II, LNAI **6635**, 51–62 (2011)
3. Chen, N., Chen, A., Zhou, L., Lu, L.: A graph-based clustering algorithm in large transaction. Intell. Data Anal. **5**(4), 327–338 (2001)
4. Guha, S., Rastogi, R., Shim, K.: ROCK: A robust clustering algorithm for categorical attributes. In: Proceedings of the 15th International Conference On Data Engeneering, pp. 512–521 (1999)
5. Huang, X., Lai, W.: Clustering graphs for visualization via node similarities. J. Vis. Lang. Comput. **17**, 225–253 (2006)
6. Karypis, G., Han, E.-H., Kumar, V.: Chameleon: Hierarchical clustering using dynamic modeling. IEEE Comput. **32**(8), 68–75 (1999)
7. Kawaji, H., Takenaka, Y., Matsuda, H.: Graph-based clustering for finding distant relationships in a large set of protein sequences. Bioinformatics **20**(2), 243–252 (2004)
8. Novák, P., Neumann, P., Macas, J.: Graph-based clustering and characterization of repetitive sequences in next-generation sequencing data. BMC Bioinformatics **11**, 378 (2010)

9. Zaki, M.J., Peters, M., Assent, I., Seidl, T.: CLICKS: An effective algorithm for mining subspace clusters in categorical datasets. Data Knowl. Eng. **60**, 51–70 (2007)

10. McQueen, J.: Some methods for classification and analysis of multivariate observations. In: Proceedings of Fifth Berkeley Symposium on Mathematical Statistics and Probability, pp. 281–297 (1967)

11. Martinetz, T.M., Shulten, K.J.: A neural-gas network learns topologies. In Kohonen, T., Mäkisara, K., Simula, O., Kangas, J. (eds): Artificial Neural Networks, pp. 397–402 (1991)

12. Kohonen, T.: Self-Organizing Maps, 3rd edn. Springer, New York (2001)

13. Bernstein, M., de Silva, V., Langford, J.C., Tenenbaum, J.B.: Graph approximations to geodesics on embedded manifolds. Stanford University (2000)

Contents

Acronyms

AHIGG	Adaptive Hierarchical Incremental Grid Growing
BMU	Best Matching Unit
CA	Cluster Analysis
CCA	Curvilinear Component Analysis
CDA	Curvilinear Distance Analysis
CHL	Competitive Hebbian Learning
DM	Data Mining
DT	Delaunay Triangulation
DTRN	Dynamic Topology Representing Network
EDA	Exploratory Data Analysis
FCM	Fuzzy c-Means
FC-WINN	Fuzzy Clustering using Weighted Incremental Neural Network
GCS	Growing Cell Structures
GNG	Growing Neural Gas algorithm
GNLP-NG	Geodesic Nonlinear Projection Neural Gas
HiGS	Hierarchical Growing Cell Structures
ICA	Independent Component Analysis
IGG	Incremental Grid Growing
LBG	Linde-Buzo-Gray algorithm
LDA	Linear Discriminant Analysis
LLE	Locally Linear Embedding
MDS	Multidimensional Scaling
MND	Mutual Neighbor Distance
NG	Neural Gas
NN	Neural Network
OVI-NG	Online Visualization Neural Gas
PCA	Principal Component Analysis
SM	Sammon Mapping
SOM	Self-Organizing Map
TRN	Topology Representing Network
VQ	Vector Quantization
WINN	Weighted Incremental Neural Network

Symbols

c	Number of the clusters
C	Set of clusters
C_i	The i-th cluster
$d_{i,j}$	The distance measure of the objects $\mathbf{x}_i$ and $\mathbf{x}_j$
D	The dimension of the observed data set
M	A manifold
N	Number of the observed objects
$s_{i,j}$	The similarity measure of the objects $\mathbf{x}_i$ and $\mathbf{x}_j$
$\mathbf{U}$	The fuzzy partition matrix
$\mu_{i,k}$	An element of the fuzzy partition matrix
$\mathbf{V}$	The set of the cluster centers
$\mathbf{v}_i$	A cluster center
$\mathbf{W}$	The set of the representatives
$\mathbf{w}_i$	A representative element (a codebook vector)
$\mathbf{X}$	The set of the observed objects
$\mathbf{x}_i$	An observed object
$\mathbf{Z}$	The set of the mapped objects
$\mathbf{z}_i$	A low-dimensional mapped object

Chapter 1
Vector Quantisation and Topology Based Graph Representation

Abstract Compact graph based representation of complex data can be used for clustering and visualisation. In this chapter we introduce basic concepts of graph theory and present approaches which may generate graphs from data. Computational complexity of clustering and visualisation algorithms can be reduced replacing original objects with their representative elements (code vectors or fingerprints) by vector quantisation. We introduce widespread vector quantisation methods, the k-means and the neural gas algorithms. Topology representing networks obtained by the modification of neural gas algorithm create graphs useful for the low-dimensional visualisation of data set. In this chapter the basic algorithm of the topology representing networks and its variants (Dynamic Topology Representing Network and Weighted Incremental Neural Network) are presented in details.

1.1 Building Graph from Data

A *graph* G is a pair (V, E), where V is a finite set of the elements, called *vertices* or *nodes*, and E is a collection of pairs of V. An element of E, called *edge* , is $e_{i,j} = (v_i, v_j)$, where $v_i, v_j \in V$. If $\{u, v\} \in E$, we say that u and v are *neighbors*. The set of the neighbors for a given vertex is the *neighborhood* of that vertex. The *complete graph* K_N on a set of N vertices is the graph that has all the $\binom{N}{2}$ possible edges. In a *weighted graph* a weight function $w : E \rightarrow \mathbb{R}$ is defined, which function determines a weight $w_{i,j}$ for each edge $e_{i,j}$. A graph may be *undirected*, meaning that there is no distinction between the two vertices associated with each edge. On the other hand, a graph may be *directed*, when its edges are directed from one vertex to another. A graph is *connected* if there is a path (i.e. a sequence of edges) from any vertex to any other vertex in the graph. A graph that is not connected is said to be *disconnected*. A graph is *finite* if V and E are finite sets. A *tree* is a graph in which any two vertices are connected by exactly one path. A *forest* is a disjoint union of trees.

Á. Vathy-Fogarassy and J. Abonyi, *Graph-Based Clustering and Data Visualization Algorithms*, SpringerBriefs in Computer Science, DOI: 10.1007/978-1-4471-5158-6_1, © János Abonyi 2013

A *path* from $v_{start} \in V$ to $v_{end} \in V$ in a graph is a sequence of edges in E starting with at vertex $v_0 = v_{start}$ and ending at vertex $v_{k+1} = v_{end}$ in the following way: $(v_{start}, v_1)(v_1, v_2), \ldots, (v_{k-1}, v_k), (v_k, v_{end})$. A *circle* is a simple path that begins and ends at the same vertex.

The distance between two vertices v_i and v_j of a finite graph is the minimum length of the paths (sum of the edges) connecting them. If no such path exists, then the distance is set equal to ∞. The distance from a vertex to itself is zero. In graph based clustering the *geodesic distance* is most frequently used concept instead of the graph distance, because of it expresses the length of the path along the structure of manifold. Shortest paths from a vertex to other vertices can be calculated by Dijkstra's algorithm, which is given in Appendix A.2.1.

Spanning trees play important role in the graph based clustering methods. Given a $G = (V, E)$ connected undirected graph. A *spanning tree* $(T = (V, E'), E' \subseteq E)$ of the graph $G = (V, E)$ is a subgraph of G that is a tree, and it connects all edges of G together. If the number of the vertices is N, then a spanning tree has exactly $N - 1$ edges. The *Minimal spanning tree* (MST) [1] of a weighted graph is a spanning tree where the sum of the edge weights is minimal. We have to mention that there may exist several different minimal spanning trees of a given graph. The minimal spanning tree of a graph can be easy constructed by Prim's or Kruskal's algorithm. These algorithms are presented in Appendices A.1.1 and A.1.2.

To build a graph that emphasises the real structure of data the intrinsic relations of data should be modelled. There are two basic approaches to connect neighbouring objects together: ε-*neighbouring* and k-*neighbouring*. In case of ε-neighbouring approach two objects $\mathbf{x}_i$ and $\mathbf{x}_j$ are connected by an edge if they are lying in an ε radius environment ($d_{i,j} < \varepsilon$, where $d_{i,j}$ yields the 'distance' of the objects $\mathbf{x}_i$ and $\mathbf{x}_j$, and ε is a small real number). Applying the k-neighbouring approach, two objects are connected to each other if one of them is in among the k-nearest neighbours of the other, where k is the number of the neighbours to be taken into account. This method results in the k nearest neighbour graph (*knn graph*). The edges of the graph can be weighted several ways. In simplest case, we can assign the Euclidean distance of the objects to the edge connecting them together. Of course, there other possibilities as well, for example the number of common neighbours can also characterise the strength of the connectivity of data.

1.2 Vector Quantisation Algorithms

In practical data mining data often contain large number of observations. In case of large datasets the computation of the complete weighted graph requires too much time and storage space. Data reduction methods may provide solution for this problem. Data reduction can be achieved in such a way that the original objects are replaced with their representative elements. Naturally, the number of the representative elements is considerably less than the number of the original observations. This form of data reduction methods is called *Vector quantization* (VQ). Formally, vector quantisation is the process of quantising D-dimensional input vectors to a reduced set of D-dimensional output vectors referred to as representatives or *codebook vectors*. The set of the codebook vectors is called codebook also referred as cluster centres or fingerprints. Vector quantisation is widely used method in many data compression applications, for example in image compression [2–4], in voice compression and identification [5–7] and in pattern recognition and data visualization [8–11].

In the following we introduce the widely used vector quantisation algorithms: k-means clustering, neural gas and growing neural gas algorithms, and topology representing networks. Except the k-means all approaches result in a graph which emphasises the dominant topology of the data. Kohonen Self-Organizing Map is also referred as a vector quantisation method, but this algorithm includes dimensionality reduction as well, so this method will be presented in Sect. 3.4.2.

1.2.1 k-Means Clustering

k-means algorithm [12] is the simplest and most commonly used vector quantisation method. k-means clustering partitions data into clusters and minimises distance between cluster centres (code vectors) and data related to the clusters:

$$J(\mathbf{X}, \mathbf{V}) = \sum_{i=1}^{c} \sum_{\mathbf{x}_k \in C_i} \|\mathbf{x}_k - \mathbf{v}_i\|^2, \tag{1.1}$$

where C_i denotes the ith cluster, and $\|\mathbf{x}_k - \mathbf{v}_i\|$ is a chosen distance measure between the data point $\mathbf{x}_k$ and the cluster center $\mathbf{v}_i$.

The whole procedure can be found in Algorithm 1.

Algorithm 1 k-means algorithm

Step 1   Choose the number of clusters, k.
Step 2   Generate k random points as cluster centers.
Step 3   Assign each point to the nearest cluster center.
Step 4   Compute the new cluster centers as the centroids of the clusters.
Step 5   If the convergence criterion is not met go back to Step 3.

The iteration steps are repeated until there is no reassignment of patterns to new cluster centers or there is no significant decrease in the squared error.

The k-means algorithm is very popular because it is easy to implement, and its time complexity is $O(N)$, where N is the number of objects. The main drawback of this algorithm is that it is sensitive to the selection of the initial partition and may converge to a local minimum of the criterion function. As its implementation is very easy, this algorithm is frequently used for vector quantisation. Cluster centres can be seen as the reduced representation (representative elements) of the data. The number of the cluster centres and so the number of the representative elements (codebook vectors) is given by the user a priori. The *Linde-buzo-gray algorithm* (LBG) [13] works similar to the k-means vector quantisation method, but it starts with only one representative element (it is the cluster centre or centroid of the entire data set) and in each iteration dynamically duplicates the number of the representative elements and reassigns the objects to be analysed among the cluster centres. The algorithm stops when the desired number of centroids is obtained.

Partitional clustering is closely related to the concept of *Voronoi diagram*. A set of representative elements (cluster centres) decompose subspaces called Voronoi cells. These Voronoi cells are drawn in such a way that all data points in a given Voronoi cell are closer to their own representative data point than to the other representative elements. *Delaunay triangulation* (DT) is the dual graph of the Voronoi diagram for the same representatives. Delaunay triangulation [14] is a subdivision of the space into triangles in such a way that there no other representative element is inside the circumcircle of any triangle. As a result the DT divides the plane into a number of triangles. Figure 1.1 represents a small example for the Voronoi diagram and Delaunay triangulation. In this figure blue dots represents the representative objects, the Voronoi cells are drawn with red lines, and black lines form the Delaunay triangulation of the representative elements. In this approach the representative elements can be seen as a compressed presentation of the space in such a way that data points placed in a Voronoi cell are replaced with their representative data point in the same Voronoi cell.

The *induced Delaunay triangulation* is a subset of the Delaunay triangulation, and it can be obtained by masking the Delaunay triangulation with the data distribution. Therefore the induced Delaunay triangulation reflects more precisely to the structure of data and do not contains such edges which go through in such areas where no data

Fig. 1.1 The Voronoi diagram and the Delaunay triangulation

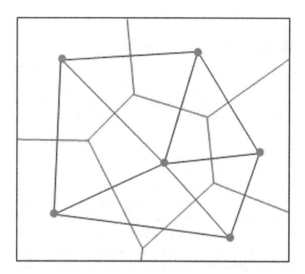

points are found. The detailed description of induced Delaunay triangulation and the connecting concept of masked Voronoi polyhedron can be found in [15].

1.2.2 Neural Gas Vector Quantisation

Neural gas algorithm (NG) [16] gives an informative reduced data representation for a given data set. The name 'neural gas' is coming from the operation of the algorithm since representative data points distribute themselves in the vector space like a gas. The algorithm firstly initialises code vectors randomly. Then it repeats iteration steps in which the following steps are performed: the algorithm randomly chooses a data point from the data objects to be visualised, calculates the distance order of the representatives to the randomly chosen data point, and in the course of the adaptation step the algorithm moves all representatives closer to the randomly chosen data point. The detailed algorithm is given in Algorithm 2.

Algorithm 2 The neural gas algorithm

Given a set of input objects $\mathbf{X} = \{\mathbf{x}_1, \mathbf{x}_2, \ldots, \mathbf{x}_N\}, \mathbf{x}_i \in \mathbb{R}^D, i = 1, 2, \ldots, N$.

Step 1 Initialize randomly all representative data points $\mathbf{w}_i \in \mathbb{R}^D, i = 1, 2, \ldots, n \ (n < N)$. Set the iteration counter to $t = 0$.

Step 2 Select an input object $(\mathbf{x}_i(t))$ with equal probability for all objects.

Step 3 Calculate the distance order for all $\mathbf{w}_j$ representative data points with respect to the selected input object $\mathbf{x}_i$. Denote j_1 the index of the closest codebook vector, j_2 the index of the second closest codebook vector and so on.

Step 4 Move closer all representative data points to the selected input object $\mathbf{x}_i$ based on the following formula:

$$\mathbf{w}_{j_k}^{(t+1)} = \mathbf{w}_{j_k}^{(t)} + \varepsilon(t) \cdot e^{-k/\lambda(t)} \cdot \left(\mathbf{x}_i - \mathbf{w}_{j_k}^{(t)} \right) \quad (1.2)$$

where ε is an adaptation step size, and λ is the neighborhood range.

Step 5 If the termination criterion not met increase the iteration counter $t = t + 1$, and go back to Step 2.

The ε and λ parameters are decreasing with time t. The adaptation step (Step 4) corresponds to a stochastic gradient descent on a given cost function. As a result the algorithm presents n D-dimensional output vectors which distribute themselves homogeneously in the input 'data cloud'.

Figure 1.2 shows a synthetic data set ('boxlinecircle') and the run of the neural gas algorithm on this data set. The original data set contains 7,100 sample data ($N = 7100$) placed in a cube, in a refracted line and in a circle (Fig. 1.2a). Data points placed in the cube contain random errors (noise). In this figure the original data points are yield with blue points and the borders of the points are illustrated with red lines. Figure 1.2b shows the initialisation of the neural gas algorithm, where the neurons were initialised in the range of the variables randomly. The number of

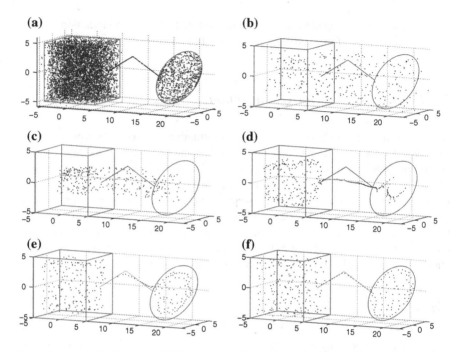

Fig. 1.2 A synthetic data set and different status of neural gas algorithm. **a** The synthetic 'box-linecircule' data set ($N = 7100$). **b** Neural gas initialization ($n = 300$). **c** NG, numbr of itrations: 100 ($n = 300$). **d** NG, number of iterations: 1000 ($n = 300$). **e** NG, number of iterations: 10000 ($n = 300$). **f** NG, number of iterations: 50000 ($n = 300$)

the representative elements was chosen to be $n = 300$. Figure 1.2c–f show different states of the neural gas algorithm. Representative elements distribute themselves homogenously and learn the form of the original data set (Fig. 1.2f).

Figure 1.3 shows an another application example. The analysed data set contains 5,000 sample points placed on a 3-dimensional S curve. The number of the representative elements in this small example was chosen to be $n = 200$, and the neurons was initialised as data points characterised by small initial values. Running results in different states are shown in Fig. 1.3b–d.

It should be noted that neural gas algorithm has much more robust convergence properties than k-means vector quantisation.

1.2.3 Growing Neural Gas Vector Quantisation

In most of the cases the distribution of high dimensional data is not known. In this cases the initialisation of the k-means and the neural gas algorithms is not easy, since it is hard to determine the number of the representative elements (clusters).

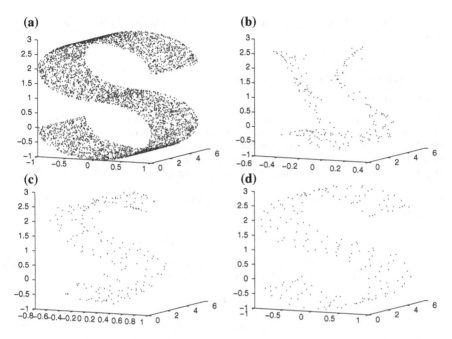

Fig. 1.3 The S curve data set and different states of the neural gas algorithm. **a** The 'S curve' data set ($N = 5000$). **b** NG, number of iterations: 200 ($n = 200$). **c** NG, number of iterations: 1000 ($n = 200$). **d** NG, number of iterations: 10000 ($n = 200$)

The *Growing neural gas* (GNG) [17] algorithm provides a fairly good solution to solve this problem, since it adds and removes representative elements dynamically. The other main benefit of this algorithm is that it creates a graph of representatives, therefore it can be used for exploring the topological structure of data as well. GNG algorithm starts with two random representatives in the vector space. After this initialisation step the growing neural gas algorithm iteratively select an input vector randomly, locate the two nearest nodes (representative elements) to this selected input vector, moves the nearest representative closer to the selected input vector, updates some edges, and in definite cases creates a new representative element as well. The algorithm is detailed in Algorithm 3 [17]. As we can see the network topology is generated incrementally during the whole process. Termination criterion might be for example the evaluation of a quality measure (or a maximum number of the nodes has been reached). GNG algorithm has several important parameters, including the maximum age of a representatives before it is deleted (a_{max}), scaling factors for the reduction of error of representatives (α, d), and the degrees (ε_b, ε_a) of movements of the selected representative elements in the adaptation step (Step 6). As these parameters are constant in time and since the algorithm is incremental, there is no need to determine the number of representatives *a priori*. One of the main benefits of growing neural gas algorithm is that is generates a graph as results. Nodes of this graph are representative elements which present the distribution of the original objects and edges give information about the neighbourhood relations of the representatives.

Algorithm 3 Growing neural gas algorithm

Given a set of input objects $\mathbf{X} = \{\mathbf{x}_1, \mathbf{x}_2, \ldots, \mathbf{x}_N\}$, $\mathbf{x}_i \in \mathbb{R}^D$, $i = 1, 2, \ldots, N$.

Step 1   Initialisation: Generate two random representatives ($\mathbf{w}_a$ and $\mathbf{w}_b$) in the D-dimensional vector space ($\mathbf{w}_a$, $\mathbf{w}_b \in \mathbb{R}^D$), and set their error variables to zero ($error(a) = 0$, $error(b) = 0$).

Step 2   Select an input data point $\mathbf{x}$ randomly, according to the data distribution.

Step 3   Find the nearest $\mathbf{w}_{s_1}$ and the second nearest $\mathbf{w}_{s_2}$ representative elements to $\mathbf{x}$.

Step 4   Increment the age of all edges emanating from the nearest representative data point $\mathbf{w}_{s_1}$ by 1.

Step 5   Update the error variable of the nearest representative element ($error(s_1)$) by adding the squared distance between $\mathbf{w}_{s_1}$ and $\mathbf{x}$ to it.

$$error(s_1) \leftarrow error(s_1) + \|\mathbf{w}_{s_1} - \mathbf{x}\|^2 \qquad (1.3)$$

Step 6   Move $\mathbf{w}_{s_1}$ and its topological neighbours (nodes connected to $\mathbf{w}_{s_1}$ by an edge) towards $\mathbf{x}$ by fractions ϵ_b and ϵ_n ($\epsilon_b, \epsilon_n \in [0, 1]$), respectively of the total distance

$$\mathbf{w}_{s_1} \leftarrow \mathbf{w}_{s_1} + \epsilon_b(\mathbf{x} - \mathbf{w}_{s_1}) \qquad (1.4)$$

$$\mathbf{w}_n \leftarrow \mathbf{w}_n + \epsilon_n(\mathbf{x} - \mathbf{w}_n) \qquad (1.5)$$

Equation 1.4 is used for the closest representative data object and Eq. 1.5 for all direct topological neighbors n of $\mathbf{w}_{s_1}$.

Step 7   If $\mathbf{w}_{s_1}$ and $\mathbf{w}_{s_2}$ are not connected then create an edge between them. If they are connected set the age of this edge to 0.

Step 8   Remove all edges with age larger than $a_{\max}$. If as a consequence of this deletion there come up nodes with no edges then remove them.

Step 9   If the current iteration is an integer multiple of a parameter λ insert a new representative element as follows:

- Find the representative element $\mathbf{w}_q$ with the largest error.
- Find the data point $\mathbf{w}_r$ with the largest error among the neighbors of $\mathbf{w}_q$.
- Insert a new representative element $\mathbf{w}_s$ halfway between the data points $\mathbf{w}_q$ and $\mathbf{w}_r$

$$\mathbf{w}_s = \frac{\mathbf{w}_q + \mathbf{w}_r}{2} \qquad (1.6)$$

- Create edges between the representatives $\mathbf{w}_s$ and $\mathbf{w}_q$, and $\mathbf{w}_s$ and $\mathbf{w}_r$. If there was an edge between $\mathbf{w}_q$ and $\mathbf{w}_r$ than delete it.
- Decrease the error variables of representatives $\mathbf{w}_q$ and $\mathbf{w}_r$, and initialize the error variable of the data point $\mathbf{w}_s$ with the new value of the error variable of $\mathbf{w}_q$ in that order as follows:

$$error(q) = \alpha \times error(q) \qquad (1.7)$$
$$error(r) = \alpha \times error(r) \qquad (1.8)$$
$$error(s) = error(q) \qquad (1.9)$$

Step 10   Decrease all error variables by multiplying them with a constant d.

Step 11   If a termination criterion is not met continue the iteration and go back to Step 2.

1.2.4 Topology Representing Network

Topology representing network (TRN) algorithm [15, 16] is one of best known neural network based vector quantisation method. The TRN algorithm works as follows. Given a set of data ($\mathbf{X} = \{\mathbf{x}_1, \mathbf{x}_2, \ldots, \mathbf{x}_N\}$, $\mathbf{x}_i \in \mathbb{R}^D$, $i = 1, \ldots, N$) and a set of codebook vectors ($\mathbf{W} = \{\mathbf{w}_1, \mathbf{w}_2, \ldots, \mathbf{w}_n\}$, $\mathbf{w}_i \in \mathbb{R}^D$, $i = 1, \ldots, n$) ($N > n$) the algorithm distributes pointers $\mathbf{w}_i$ between the data objects by the neural gas algorithm (steps 1–4 without setting the connection strengths $c_{i,j}$ to zero) [16], and forms connections between them by applying competitive Hebbian rule [18]. The run of the algorithm results in a Topology Representing Network that means a graph $G = (\mathbf{W}, \mathbf{C})$, where $\mathbf{W}$ denotes the nodes (codebook vectors, neural units, representatives) and $\mathbf{C}$ yields the set of edges between them. The detailed description of the TRN algorithm is given in Algorithm 4.

Algorithm 4 TRN algorithm

Given a set of input objects $\mathbf{X} = \{\mathbf{x}_1, \mathbf{x}_2, \ldots, \mathbf{x}_N\}$, $\mathbf{x}_i \in \mathbb{R}^D$, $i = 1, 2, \ldots, N$.

Step 1 Initialise the codebook vectors $\mathbf{w}_j$ ($j = 1, \ldots, n$) randomly. Set all connection strengths $c_{i,j}$ to zero. Set $t = 0$.

Step 2 Select an input pattern $\mathbf{x}_i(t)$, ($i = 1, \ldots, N$) with equal probability for each $\mathbf{x} \in \mathbf{X}$.

Step 3 Determine the ranking $r_{i,j} = r(\mathbf{x}_i(t), \mathbf{w}_j(t)) \in \{0, 1, \ldots, n - 1\}$ for each codebook vector $\mathbf{w}_j(t)$ with respect to the vector $\mathbf{x}_i(t)$ by determining the sequence $(j_0, j_1, \ldots, j_{n-1})$ with

$$\|\mathbf{x}_i(t) - \mathbf{w}_{j_0}(t)\| < \|\mathbf{x}_i(t) - \mathbf{w}_{j_1}(t)\| < \cdots < \|\mathbf{x}_i(t) - \mathbf{w}_{j_{n-1}}(t)\|. \tag{1.10}$$

Step 4 Update the codebook vectors $\mathbf{w}_j(t)$ according to the neural gas algorithm by setting

$$\mathbf{w}_j(t + 1) = \mathbf{w}_j(t) + \varepsilon \cdot e^{-\frac{r(\mathbf{x}_i(t), \mathbf{w}_j(t))}{\lambda(t)}} \left(\mathbf{x}_i(t) - \mathbf{w}_j(t)\right), \quad j = 1, \ldots, n \tag{1.11}$$

Step 5 If a connection between the first and the second closest codebook vector to $\mathbf{x}_i(t)$ does not exist already ($c_{j_0, j_1} = 0$), create a connection between them by setting $c_{j_0, j_1} = 1$ and set the age of this connection to zero by $t_{j_0, j_1} = 0$. If this connection already exists ($c_{j_0, j_1} = 1$), set $t_{j_0, j_1} = 0$, that is, refresh the connection of the codebook vectors $j_0 - j_1$.

Step 6 Increment the age of all connections of $\mathbf{w}_{j_0}(t)$ by setting $t_{j_0, l} = t_{j_0, l} + 1$ for all $\mathbf{w}_l(t)$ with $c_{j_0, l} = 1$.

Step 7 Remove those connections of codebook vector $\mathbf{w}_{j_0}(t)$ the age of which exceed the parameter T by setting $c_{j_0, l} = 0$ for all $\mathbf{w}_l(t)$ with $c_{j_0, l} = 1$ and $t_{j_0, l} > T$.

Step 8 Increase the iteration counter $t = t + 1$. If $t < t_{\max}$ go back to Step 2.

The algorithm has many parameters. Opposite to growing neural gas algorithm topology representing network requires the number of the representative elements a priori. The number of the iterations ($t_{\max}$) and the number of the codebook vectors (n) are determined by the user. Parameter λ, step size ε and lifetime T are dependent on the number of the iterations. This time dependence can be expressed by the following general form:

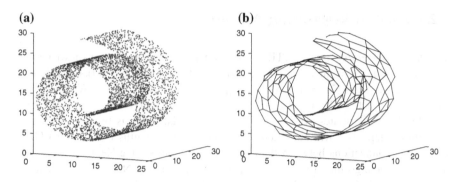

Fig. 1.4 The swiss roll data set and a possible topology representing network of it. **a** Original swiss roll data set ($N = 5000$). **b** TRN of swiss roll dats set ($n = 200$)

$$g(t) = g_i \left(\frac{g_f}{g_i} \right)^{t/t_{\max}} , \qquad (1.12)$$

where g_i denotes the initial value of the variable, g_f denotes the final value of the variable, t denotes the iteration number, and $t_{\max}$ denotes the maximum number of iterations. (For example for parameter λ it means: $\lambda(t) = \lambda_i (\lambda_f/\lambda_i)^{t/t_{\max}}$.) Paper [15] gives good suggestions to tune these parameters.

To demonstrate the operation of TRN algorithm 2 synthetic data sets were chosen. The swiss roll and the S curve data sets. The number of original objects in both cases were $N = 5000$. The swiss roll data set and its topology representing network with $n = 200$ quantised objects are shown in Fig. 1.4a and b.

Figure 1.5 shows two possible topology representing networks of the S curve data set. In Fig. 1.5a, a possible TRN graph of the S curve data set with $n = 100$ representative elements is shown. In the second case (Fig. 1.5b) the number of the representative elements was chosen to be twice as many as in the first case. As it can

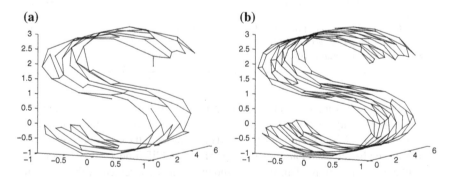

Fig. 1.5 Different topology representing networks of the S curve data set. **a** TRN of S curve data set ($n = 100$). **b** TRN of S curve dats set ($n = 200$)

be seen the greater the number of the representative elements the more accurate the approximation is.

Parameters in both cases were set as follows: the number of iterations was set to $t_{\max} = 200n$, where n is the number of representative elements. Initial and final values of λ, ε and T parameters were: $\varepsilon_i = 0.3$, $\varepsilon_f = 0.05$, $\lambda_i = 0.2n$, $\lambda_i = 0.01$, $T_i = 0.1n$ and $T_i = 0.05n$. Although the modification of these parameters may somewhat change the resulted graph, the number of the representative elements has more significant effect on the structure of the resulted network.

1.2.5 Dynamic Topology Representing Network

The main disadvantage of the TRN algorithm is that the number of the representatives must be given a priori. The *Dynamic topology representing network* (DTRN) introduced by Si et al. in 2000 [19] eliminates this drawback. In this method the graph incrementally changes by adding and removing edges and vertices. The algorithm starts with only one node, and it examines a vigilance test in each iteration. If the nearest (winner) node to the randomly selected input pattern fails this test, a new node is created and this new node is connected to the winner. If the winner passes the vigilance test, the winner and its adjacent neighboors are moved closer to the selected input pattern. In this second case, if the winner and the second closest nodes are not connected, the algorithm creates an edge between them. Similarly to the TRN algorithm DTRN also removes those connections whose age achieves a predefined threshold. The most important input parameter of DTRN algorithm is the vigilance threshold. This vigilance threshold gradually decreases from an initial value to a final value. The detailed algorithm is given in Algorithm 5.

The termination criterion of the algorithm can be given by a maximum number of iterations or can be controlled with the vigilance threshold. The output of the algorithm is a D-dimensional graph.

As it can be seen DTRN and TRN algorithms are very similar to each other, but there are some significant differences between them. While TRN starts with n randomly generated codebook vectors, DTRN step by step builds up the set of the representative data elements, and the final number of the codebook vectors can be determined by the vigilance threshold as well. While during the adaptation process the TRN moves the representative elements based on their ranking order closer to the selected input object, DTRN performs this adaptation step based on the Euclidean distances of the representatives and the selected input element. Furthermore, TRN moves all representative elements closer to the selected input object, but DTRN method applies the adaptation rule only to the winner and its direct topological neighboors. The vigilance threshold is an additional parameter of the DTRN algorithm. The tuning of this is based on the formula introduced in the TRN algorithm. The vigilance threshold ρ accordingly to the formula 1.12 gradually decreases from ρ_i to ρ_f during the algorithm.

Algorithm 5 DTRN algorithm

Given a set of input objects $\mathbf{X} = \{\mathbf{x}_1, \mathbf{x}_2, \ldots, \mathbf{x}_N\}$, $\mathbf{x}_i \in \mathbb{R}^D$, $i = 1, 2, \ldots, N$.

Step 1   Initialization: Start with only one representative element (node) $\mathbf{w}_i$. To represent this node select one input object randomly.

Step 2   Select randomly an element $\mathbf{x}$ from the input data objects. Find the nearest representative element (the winner) ($\mathbf{w}_c$) and its direct neighbor ($\mathbf{w}_d$) from:

$$\|\mathbf{x} - \mathbf{w}_c\| = \min_i \|\mathbf{x} - \mathbf{w}_i\| \tag{1.13}$$

$$\|\mathbf{x} - \mathbf{w}_d\| = \min_{i \neq c} \|\mathbf{x} - \mathbf{w}_i\| \tag{1.14}$$

The similarity between a data point and a representative element is measured by the Euclidean distance.

Step 3   Perform a vigilance test based on the following formula:

$$\|\mathbf{x} - \mathbf{w}_c\| < \rho \tag{1.15}$$

where ρ is a vigilance threshold.

Step 4   If the winner representative element fails the vigilance test: create a new codebook vector with $\mathbf{w}_g = \mathbf{x}$. Connect the new codebook vector to the winner representative element by setting $s_{c,g} = 1$, and set other possible connections of $\mathbf{w}_g$ to zero. Set $t_{g,j} = 0$ if $j = c$ and $t_{g,j} = \infty$ otherwise. Go Step 6.

Step 5   If the winner representative element passes the vigilance test:

Step 5.1: Update the coordinates of the winner node and its adjacent neighbors based on the following formula:

$$\Delta\mathbf{w}_i(k) = s_{c,i}\alpha(k)\frac{e^{-\beta\|\mathbf{x}(k)-\mathbf{w}_i(k)\|^2}}{\sum_{j=1}^{L} s_{c,j}e^{-\beta\|\mathbf{x}(k)-\mathbf{w}_i(k)\|^2}}\left(\mathbf{x}(k) - \mathbf{w}_i(k)\right), \quad i = 1, \ldots, L, \tag{1.16}$$

where $k = 0, 1, \ldots$ is a discrete time variable, $\alpha(k)$ is the learning rate factor, and β is an annealing parameter.

Step 5.2: Update the connections between the representative elements. If the winner and its closest representative are connected ($s_{c,d} = 1$) set $t_{c,d} = 0$. If they are not connected with an edge, connect them by setting $s_{c,d} = 1$ and set $t_{c,d} = 0$.

Step 6   Increase all connections to the winner representative element by setting $t_{c,j} = t_{c,j} + 1$. If an age of a connection exceeds a time limit T ($t_{c,j} > T$) delete this edge by setting $s_{c,j} = 0$.

Step 7   Remove the node $\mathbf{w}_i$ if $s_{i,j} = 0$ for all $j \neq i$, and there exists more than 1 representative element. That is if there are more than 1 representative elements, remove all representatives which do not have any connections to the other codebook vectors.

Step 8   If a termination criterion is not met continue the iteration and go back to Step 2.

As a result, DTRN overcomes the difficulty of TRN by applying the vigilance threshold, however the growing process is still determined by a user defined threshold value. Furthermore both algorithms have difficulty breaking links between two separated areas.

Next examples demonstrate DTRN and the effect of parametrisation on the resulted graph. Figure 1.6 shows two possible results on the S curve data set. The

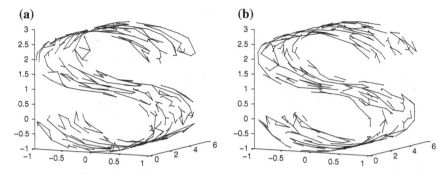

Fig. 1.6 Different DTRN graphs of the S curve data set with the same parameter settings. **a** A possible DTRN of S curve data set ($n = 362$). **b** Another possible DTRN of S curve data set ($n = 370$)

algorithm in these two cases was parameterised in the same way as follows: the vigilance threshold decreased from the average deviation of the dimensions to constant 0.1, learning rate factor decreased from 0.05 to 0.0005, number of the iterations was chosen to be 1,000 and maximum age of connections was set to be 5. DTRN results in different topology based networks arising from the random initialisation of the neurons. As DTRN dynamically adds and removes nodes the number of the representative elements differs in the two examples.

Figure 1.7 shows the influence of the number of iterations (t_{max}) and the maximum age (T) of edges. When the number of the iterations increases the number of representative elements increases as well. Furthermore, the increase of the maximum age of edges results additional links between slightly far nodes (see Fig. 1.7b and d).

1.2.6 Weighted Incremental Neural Network

H.H. Muhammad proposed an extension of the TRN algorithm, called *Weighted incremental neural network* (WINN) [20]. This algorithm can be seen as a modified version of the Growing Neural Gas algorithm. The Weighted Incremental Neural Network method is based on neural network approach as it produces a weighted connected net. The resulted graph contains weighted edges connected by weighted nodes where weights are proportional to the local densities of the data.

The algorithm starts with two randomly selected nodes from the data. In each iteration the algorithm selects one additional object and the nearest node to this object and its direct topological neighboors are moved towards this selected object. When the nearest node and the other $n - 1$ nearest nodes are not connected the algorithm establishes a connection between them. The ages and the weight-variables of edges, the error-variables and the weights of nodes are updated step by step. This method inserts a new node to the graph when the number of the generated input pattern is

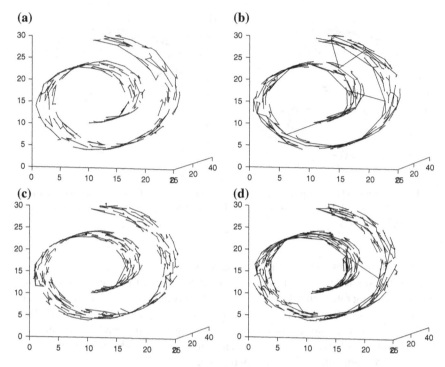

Fig. 1.7 DTRN graphs of the swiss roll data set with different parameter settings. **a** DTRN of swiss roll data set $t_{max} = 500, T = 5 (n = 370)$. **b** DTRN of swiss roll data set $t_{max} = 500, T = 10$ $(n = 383)$. **c** DTRN of swiss roll data set $t_{max} = 1000, T = 5 (n = 631)$. **d** DTRN of swiss roll data set $t_{max} = 1000, T = 10 (n = 345)$

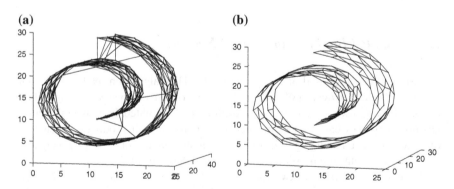

Fig. 1.8 Weighted Incremental Networks of the swiss roll data set. **a** WINN of swiss roll data set applying the suggested $a_{max} = N/10$ parameter setting. **b** WINN of swiss roll data set with $T = 3$ parameter setting

a multiple of a predefined λ parameter. Similarly to the previous algorithms WINN also removes the 'old' connections. The whole algorithm is given in Algorithm 6.

The algorithm has several parameters. While some of them (a_{max}, λ) are dependent on the number of objects to be analysed, others (ε_b, ε_n, α and d) are independent form the size of the dataset. It is suggesed to initialise these independent parameters as follows: $\varepsilon_b = 0.05$, $\varepsilon_n = 0.0006$, $\alpha = 0.5$, and $d = 0.0005$. Parameters a_{max} and λ influence the resulted number of nodes in the graph. These parameters are suggested to set as follows: $a_{max} = $ *number of input data objects*/10, and $\lambda = $ *number of input signals that must be generated / desired number of representative elements*. The main disadvantage of the Weighted Incremental Neural Network algorithm is the difficulty of tuning these parameters.

In the course of our tests we have found that the suggested setting of parameter a_{max} is too high. In our experimental results in case of linear manifolds nonlinearly embedded in higher dimensional space lower values of parameter a_{max} gave better results. Figure 1.8a shows WINN on swiss roll data set with $N = 5000$. Following

---

Algorithm 6 WINN algorithm

Given a set of input objects $\mathbf{X} = \{\mathbf{x}_1, \mathbf{x}_2, \ldots, \mathbf{x}_N\}$, $\mathbf{x}_i \in \mathbb{R}^D$, $i = 1, 2, \ldots, N$.

Step 1 Initialization: Set the weight and the error variables of the objects to 0.

Step 2 Select randomly two nodes from the input data set $\mathbf{X}$.

Step 3 Select randomly an element (input signal) $\mathbf{x}_s$ from the input data objects.

Step 4 Find the n nearest input objects $\mathbf{x}_j$ to $\mathbf{x}_s$. Yield the first nearest object $\mathbf{x}_1$, the second nearest object $\mathbf{x}_2$, and so on. Increment the weight of n nearest objects by 1.

Step 5 Increment the age variable of all edges connected to $\mathbf{x}_1$ by 1.

Step 6 Update the error variable of $\mathbf{x}_1$ as follows:

$$\Delta err(\mathbf{x}_1) = \|\mathbf{x}_1 - \mathbf{x}_s\|^2 \qquad (1.17)$$

Step 7 Move the nearest object $\mathbf{x}_1$ and the objects connected to $\mathbf{x}_1$ towards $\mathbf{x}_s$ by fractions ε_b and ε_n respectively of their distances to $\mathbf{x}_s$.

Step 8 If there is not edges between $\mathbf{x}_s$ and $\mathbf{x}_j$ ($j = 1, 2, \ldots n$) create them, and set their age variable to 0. If these edges (or some of them) exist refresh them by setting their age variable to zero. Increment the weight variable of edges between $\mathbf{x}_s$ and $\mathbf{x}_j$ ($j = 1, 2, \ldots n$) by 1.

Step 9 Remove the edges with age more than a predefined parameter a_{max}. Isolated data points, which are not connected by any edge are also deleted.

Step 10 If the number of the generated input signals so far is multiple of a user defined parameter λ, insert a new node as follows: Determine the node $\mathbf{x}_q$ with the largest accumulated error.

Step 10.1 Insert a new node ($\mathbf{x}_r$) halfway between $\mathbf{x}_q$ and its neighbor $\mathbf{x}_f$ with the largest error. Set the weight variable of $\mathbf{x}_r$ to the average weights of $\mathbf{x}_q$ and $\mathbf{x}_f$.

Step 10.2 Connect $\mathbf{x}_r$ to $\mathbf{x}_q$ and $\mathbf{x}_f$. Initialize the weight variable of the new edges with the weight variable of edge between $\mathbf{x}_q$ and $\mathbf{x}_f$. Delete the edge connecting the nodes $\mathbf{x}_q$ and $\mathbf{x}_f$.

Step 10.3 Decrease the error variable of $\mathbf{x}_q$ and $\mathbf{x}_f$ by multiplying them with a constant α. Set the error variable of the new node $\mathbf{x}_r$ to the new error variable of $\mathbf{x}_q$.

Step 11 Decrease all error variables by multiplying them with a constant d.

Step 12 If a termination criterion not met go back to Step 3.

---

the instructions of [20] we have set parameter a_{max} to be $N/10$, $a_{max} = 500$. The resulted graph contains some unnecessary links. Setting this parameter to a lower value this superfluous connections do not appear in the graph. Figure 1.8b shows this reduced parameter setting, where a_{max} was set to be $a_{max} = 3$. The number of representative elements in both cases was $n = 200$.

References

1. Yao, A.: On constructing minimum spanning trees in k-dimensional spaces and related problems. SIAM J. Comput. 721–736 (1892)
2. Boopathy, G., Arockiasamy, S.: Implementation of vector quantization for image compression—a survey. Global J. Comput. Sci. Technol. 10(3), 22–28 (2010)
3. Domingo, F., Saloma, C.A.: Image compression by vector quantization with noniterative derivation of a codebook: applications to video and confocal images. Appl. Opt. 38(17), 3735–3744 (1999)
4. Garcia, C., Tziritas, G.: Face detection using quantized skin color regions merging and wavelet packet analysis. IEEE Trans. Multimedia 1(3), 264–277 (1999)
5. Biatov, K.: A high speed unsupervised speaker retrieval using vector quantization and second-order statistics. CoRR Vol. abs/1008.4658 (2010)
6. Chu, W.C.: Vector quantization of harmonic magnitudes in speech coding applications a survey and new technique. EURASIP J. App. Sig. Proces. 17, 2601–2613 (2004)
7. Kekre, H.B., Kulkarni, V.: Speaker identification by using vector quantization. Int. J. Eng. Sci. Technol. 2(5), 1325–1331 (2010)
8. Abdelwahab, A.A., Muharram, N.S.: A fast codebook design algorithm based on a fuzzy clustering methodology. Int. J. Image Graph. 7(2), 291–302 (2007)
9. Kohonen, T.: Self-Organizing Maps, 3rd edn. Springer, New York (2001)
10. Kurasova, O., Molyte, A.: Combination of vector quantization and visualization. Lect. Notes Artif. Intell. 5632, 29–43 (2009)
11. Vathy-Fogarassy, A., Kiss, A., Abonyi, J.: Topology representing network map—a new tool for visualization of high-dimensional data. Trans. Comput. Sci. I 4750, 61–84 (2008)
12. McQueen, J.: Some methods for classification and analysis of multivariate observations. In: Proceedings of Fifth Berkeley Symposium on Mathematical Statistics and Probability, pp. 281–297 (1967)
13. Linde, Y., Buzo, A., Gray, R.M.: An algorithm for vector quantizer design. IEEE Trans. Commun. 28, 84–94 (1980)
14. Delaunay, B.: Sur la sphere vide. Izvestia Akademii Nauk SSSR, Otdelenie Matematicheskikh i Estestvennykh Nauk 7, 793–800 (1934)
15. Martinetz, T.M., Shulten, K.J.: Topology representing networks. Neural Netw. 7(3), 507–522 (1994)
16. Martinetz, T.M., Shulten, K.J.: A neural-gas network learns topologies. In Kohonen, T., Mäkisara, K., Simula, O., Kangas, J. (eds.) Artificial Neural Networks, pp. 397–402, Elsevier Science Publishers B.V, North-Holland (1991)
17. Fritzke, B.: A growing neural gas network learns topologies. Adv. Neural Inf. Proces. Syst. 7, 625–632 (1995)
18. Hebb, D.O.: The Organization of Behavior. John, Inc New York (1949)
19. Si, J., Lin, S., Vuong, M.-A.: Dynamic topology representing networks. Neural Netw. 13, 617–627 (2000)
20. Muhammed, H.H.: Unsupervised fuzzy clustering using weighted incremental neural networks. Int. J. Neural Syst. 14(6), 355–371 (2004)

Chapter 2
Graph-Based Clustering Algorithms

Abstract The way how graph-based clustering algorithms utilize graphs for partitioning data is very various. In this chapter, two approaches are presented. The first hierarchical clustering algorithm combines minimal spanning trees and Gath-Geva fuzzy clustering. The second algorithm utilizes a neighborhood-based fuzzy similarity measure to improve k-nearest neighbor graph based Jarvis-Patrick clustering.

2.1 Neigborhood-Graph-Based Clustering

Since clustering groups neighboring objects into same cluster neighborhood graphs are ideal for cluster analysis. A general introduction to the neighborhood graphs is given in [18]. Different interpretations of concepts 'near' or 'neighbour' lead to a variety of related graphs. The *Nearest Neighbor Graph* (NNG) [9] links each vertex to its nearest neighbor. The *Minimal Spanning Tree* (MST) [29] of a weighted graph is a spanning tree where the sum of the edge weights is minimal. The *Relative Neighborhood Graph* (RNG) [25] connects two objects if and only if there is no other object that is closer to both objects than they are to each other. In the *Gabriel Graph* (GabG) [12] two objects, p and q, are connected by an edge if and only if the circle with diameter pq does not contain any other object in its interior. All these graphs are subgraphs of the well-known *Delaunay triangulation* (DT) [11] as follows:

$$NNG \subseteq MST \subseteq RNG \subseteq GabG \subseteq DT \tag{2.1}$$

There are many graph-based clustering algorithms that utilize neighborhood relationships. Most widely known graph-theory based clustering algorithms (ROCK [16] and Chameleon [20]) also utilize these concepts. Minimal spanning trees [29] for clustering was initially proposed by Zahn [30]. Clusters arising from single linkage hierarchical clustering methods are subgraphs of the minimum spanning tree of the data [15]. Clusters arising from complete linkage hierarchical clustering methods are

Á. Vathy-Fogarassy and J. Abonyi, *Graph-Based Clustering and Data Visualization Algorithms*, SpringerBriefs in Computer Science, DOI: 10.1007/978-1-4471-5158-6_2, © János Abonyi 2013

maximal complete subgraphs, and are related to the node colorability of graphs [3]. In [2, 24], the maximal complete subgraph was considered to be the strictest definition of the clusters. Several graph-based divisive clustering algorithms are based on MST [4, 10, 14, 22, 26]. The approach presented in [1] utilizes several neighborhood graphs to find the groups of objects. Jarvis and Patrick [19] extended the nearest neighbor graph with the concept of the shared nearest neighbors. In [7] Doman et al. iteratively utilize Jarvis-Patrick algorithm for creating crisp clusters and then they fuzzify the previously calculated clusters. In [17], a node structural metric has been chosen making use of the number of shared edges.

In the following, we introduce the details and improvements of MST and Jarvis-Patrick clustering algorithms.

2.2 Minimal Spanning Tree Based Clustering

Minimal spanning tree is a weighted connected graph, where the sum of the weights is minimal. Denote $G = (V, E)$ a graph. Creating the minimal spanning tree means, that we are searching the $G' = (V, E')$, the connected subgraph of G, where $E' \subset E$ and the cost is minimal. The cost is computed in the following way:

$$\sum_{e \in E'} w(e) \tag{2.2}$$

where $w(e)$ denotes the weight of the edge $e \in E$. In a graph G, where the number of the vertices is N, MST has exactly $N - 1$ edges.

A minimal spanning tree can be efficiently computed in $O(N^2)$ time using either Prim's [23] or Kruskal's [21] algorithm. *Prim's algorithm* starts with an arbitrary vertex as the root of a partial tree. In each step of the algorithm, the partial tree grows by iteratively adding an unconnected vertex to it using the lowest cost edge, until no unconnected vertex remains. *Kruskal's algorithm* begins with the connection of the two nearest objects. In each step, the minimal pairwise distance that connects separate trees is selected, and these two trees are connected along these objects. So the Kruskal's algorithm iteratively merges two trees (or a tree with a single object) in the current forest into a new tree. The algorithm continues until a single tree remains only, connecting all points. Detailed description of these algorithms are given in Appendix A.1.1.1 and A.1.1.2.

Clustering based on minimal spanning tree is a hierarchical divisive procedure. Removing edges from the MST leads to a collection of connected subgraphs of G, which can be considered as clusters. Since MST has only $N - 1$ edges, we can choose inconsistent edge (or edges) by revising only $N - 1$ values. Using MST for clustering, we are interested in finding edges, whose elimination leads to best clustering result. Such edges are called *inconsistent edges*.

The basic idea of *Zahn's algorithm* [30] is to detect inherent separations in the data by deleting edges from the MST which are significantly longer than other edges.

Step 1	Construct the minimal spanning tree so that the edges weights are the distances between the data points.
Step 2	Remove the inconsistent edges to get a set of connected components (clusters).
Step 3	Repeat Step 2 until a terminating criterion is not satisfied.

Zahn proposed the following criterion to determine the inconsistent edges: an edge is inconsistent if its length is more than f times the average length of the edges, or more than f times the average of the length of nearby edges. This algorithm is able to detect clusters of various shapes and sizes; however, the algorithm cannot detect clusters with different densities.

Identification of inconsistent edges causes problems in the MST based clustering algorithms. Elimination of k edges from a minimal spanning tree results in $k + 1$ disconnected subtrees. In the simplest recursive theories $k = 1$. Denote δ the length of the deleted edge, and let $\mathbf{V}_1$, $\mathbf{V}_2$ be the sets of the points in the resulting two clusters. In the set of clusters, we can state that there are no pairs of points $(\mathbf{x}_1, \mathbf{x}_2), \mathbf{x}_1 \in \mathbf{V}_1, \mathbf{x}_2 \in \mathbf{V}_2$ such that $d(\mathbf{x}_1, \mathbf{x}_2) < \delta$. There are several ways to define the distance between two disconnected groups of individual objects (minimum distance, maximum distance, average distance, distance of centroids, etc.). Defining the separation between $\mathbf{V}_1$ and $\mathbf{V}_2$, we have the result that the separation is at least δ. The determination of the value of δ is very difficult because data can contain clusters with different densities, shapes, volumes, and furthermore they can also contain bridges (chain links) between the clusters. A terminating criterion determining the stop of the algorithm should be also defined.

The simplest way to delete edges from MST is based on distances between vertices. By deleting the longest edge in each iteration step we get a nested sequence of subgraphs. Several ways are known to stop the algorithm, for example the user can define the number of clusters or give a threshold value on the length, as well. Zahn suggested a global threshold value for the cutting, which considers the distribution of the data in the feature space. In [30], this threshold (δ) is based on the average weight (distances) of the MST (*Criterion-1*):

$$\delta = \lambda \frac{1}{N - 1} \sum_{e \in E'} w(e) \tag{2.3}$$

where λ is a user defined parameter, N is the number of the objects, and E' yields the set of the edges of MST. Of course, λ can be defined several ways.

Long edges of MST do not always indicate outliers or cluster separation. In case of clusters with different densities, recursive cutting of longest edges does not give the expected clustering result (see Fig. 2.1). Solving this problem Zahn [30] suggested that an edge is inconsistent if its length is at least f times as long as the average of the length of nearby edges (*Criterion-2*). Another usage of Criterion-2 based MST clustering is finding dense clusters embedded in a sparse set of points.

The first two splitting criteria are based on distance between the resulted clusters. Clusters chained by a bridge of small set of data cannot be separated by

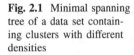

Fig. 2.1 Minimal spanning tree of a data set containing clusters with different densities

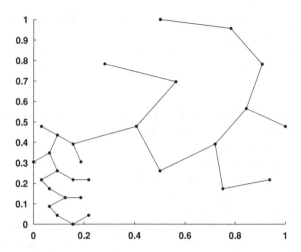

distance-based approaches (see Appendix A.6.9). To solve this chaining problem, we present a criterion based on cluster validity measure.

Many approaches use *validity measures* to assess the goodness of the obtained partitions and to estimate the correct number of clusters. This can be done in two ways:

- The first approach defines a validity function which *evaluates a complete partition*. An upper bound for the number of clusters must be estimated (c_{max}), and the algorithms have to be run with each $c \in \{2, 3, \ldots, c_{max}\}$. For each partition, the validity function provides a value such that the results of the analysis can be compared indirectly.
- The second approach consists of the definition of a validity function that *evaluates individual clusters* of a cluster partition. Again, c_{max} has to be estimated and the cluster analysis has to be carried out for c_{max}. The resulting clusters are compared to each other on the basis of the validity function. Similar clusters are collected in one cluster, very bad clusters are eliminated, so the number of clusters is reduced. The procedure can be repeated while there are clusters that do not satisfy the predefined criterion.

Different scalar validity measures have been proposed in the literature, but none of them is perfect on its own. For example, *partition index* [5] is the ratio of the sum of compactness and separation of the clusters. Compactness of a cluster means that members of the cluster should be as close to each other as possible. A common measure of compactness is the variance, which should be minimized. Separation of clusters can be measured for example based on the single linkage, average linkage approach or with the comparison of the centroid of the clusters. *Separation index* [5] uses a minimum distance separation for partition validity. *Dunn's index* [8] is originally proposed to be used at the identification of compact and well-separated clusters. This index combines the dissimilarity between clusters and their diameters to estimate the most reliable number of clusters. The problems of Dunn index are:

(i) its considerable time complexity, (ii) its sensitivity to the presence of noise in data. Three indices, are proposed in the literature that are more robust to the presence of noise. These *Dunn-like indices* are based on the following concepts: minimum spanning tree, Relative Neighborhood Graph, and Gabriel Graph.

One of the three Dunn-like indices [6] is defined using the concept of the MST. Let C_i be a cluster and $G_i = (V_i, E_i)$ the complete graph whose vertices correspond to the objects of C_i. Denote $w(e)$ the weight of an edge e of the graph. Let E_i^{MST} be the set of edges of the MST of the graph G_i, and e_i^{MST} the continuous sequence of the edges in E_i^{MST} whose total edge weight is the largest. Then, the diameter of the cluster C_i is defined as the weight of e_i^{MST}. With the use of this notation the Dunn-like index based on the concept of the MST is given by the equation:

$$D_{n_c} = \min_{i=1,\dots,n_c} \left\{ \min_{j=i+1,\dots,n_c} \left(\frac{\delta(C_i, C_j)}{\max_{k=1,\dots,n_c} \text{diam}\,(C_k)} \right) \right\} \qquad (2.4)$$

where n_c yields the number of the clusters, $\delta(C_i, C_j)$ is the dissimilarity function between two clusters C_i and C_j defined as $\min_{\mathbf{x}_l \in C_i, \mathbf{x}_m \in C_j} d(\mathbf{x}_l, \mathbf{x}_m)$, and $diam(C_k)$ is the diameter of the cluster C_k, which may be considered as a measure of clusters dispersion. The number of clusters at which D_{n_c} takes its maximum value indicates the number of clusters in the underlying data.

Varma and Simon [26] used the *Fukuyama-Sugeno clustering measure* for deleting edges from the MST. In this validity measure weighted membership value of an object is multiplied by the difference between the distance between the node and its cluster center, and the distances between the cluster center and the center of the whole data set. The Fukuyama-Sugeno clustering measure is defined in the following way:

$$FS_m = \sum_{j=1}^{N} \sum_{i=1}^{n_c} \mu_{i,j}^m \left(\|\mathbf{x}_j - \mathbf{v}_i\|_A^2 - \|\mathbf{v}_i - \mathbf{v}\|_A^2 \right) \qquad (2.5)$$

where $\mu_{i,j}$ is the degree of the membership of data point $\mathbf{x}_j$ in the ith cluster, m is a weighting parameter, $\mathbf{v}$ denotes the global mean of all objects, $\mathbf{v}_i$ denotes the mean of the objects in the ith cluster, $\mathbf{A}$ is a symmetric and positive definite matrix, and n_c denotes the number of the clusters. The first term inside the brackets measures the compactness of clusters, while the second one measures the distances of the cluster representatives. Small FS indicates tight clusters with large separations between them. Varma and Simon found, that Fukuyama-Sugeno measure gives the best performance in a data set with a large number of noisy features.

2.2.1 Hybrid MST: Gath-Geva Clustering Algorithm

In previous section, we presented main properties of minimal spanning tree based clustering methods. In the following, a new splitting method and a new clustering algorithm will be introduced.

Hybrid Minimal Spanning Tree—Gath-Geva algorithm clustering algorithm (Hybrid MST-GG) [27] first creates minimal spanning tree of the objects, then iteratively eliminates inconsistent edges and uses the resulted clusters to initialize Gaussian mixture model-based clustering algorithm (details of the Gath-Geva algorithm are given in Appendix A.5). Since clusters of MST will be approximated by multivariate Gaussians, the distribution of data can be expressed by covariance matrices of the clusters. Therefore, the proposed Hybrid MST-GG algorithm utilizes a validity measure expressed as the determinants of the covariance matrices used to represent the clusters.

The *fuzzy hyper volume* [13] validity measure is based on the concepts of hyper volume. Let $\mathbf{F}_i$ be the fuzzy covariance matrix of the ith cluster defined as

$$\mathbf{F}_i = \frac{\sum_{j=1}^{N}(\mu_{i,j})^m \left(\mathbf{x}_j - \mathbf{v}_i\right)\left(\mathbf{x}_j - \mathbf{v}_i\right)^T}{\sum_{j=1}^{N}(\mu_{i,j})^m}, \tag{2.6}$$

where $\mu_{i,j}$ denotes the degree of membership of $\mathbf{x}_j$ in cluster C_i, and $\mathbf{v}_i$ denotes the center of the ith cluster. The symbol m is the fuzzifier parameter of the fuzzy clustering algorithm and indicates the fuzzyness of clustering result. We have to mention that if the clustering result is coming from a hard clustering, the values of $\mu_{i,j}$ are either 0 or 1, and the value of m is supposed to be 1. The fuzzy hyper volume of ith cluster is given by the equation:

$$V_i = \sqrt{\det\left(\mathbf{F}_i\right)} \tag{2.7}$$

The *total fuzzy hyper volume* is given by the equation:

$$\text{FHV} = \sum_{i=1}^{c} V_i \tag{2.8}$$

where c denotes the number of clusters. Based on this measure, the proposed Hybrid Minimal Spanning Tree—Gath-Geva algorithm compares the volume of the clusters. Bad clusters with large volumes are further partitioned until there are 'bad' clusters.

In the first step, the algorithm creates the minimal spanning tree of the normalized data that will be partitioned based on the following steps:

• classical cutting criteria of the MST (Criterion-1 and Criterion-2),
• the application of fuzzy hyper volume validity measure to eliminate edges from the MST (Criterion-3).

The proposed Hybrid MST-GG algorithm iteratively builds the possible clusters. First all objects form a single cluster, and then in each iteration step a binary splitting is performed. The use of the cutting criteria results in a hierarchical tree of clusters, in which the nodes denote partitions of the objects. To refine the partitions evolved in the previous step, we need to calculate the volumes of the obtained clusters. In each iteration step, the cluster (a leaf of the binary tree) having the largest hyper volume is selected for the cutting. For the elimination of edges from the selected cluster, first the cutting conditions Criterion-1 and Criterion-2 are applied, which were previously introduced (see Sect. 2.2). The use of the classical MST based clustering methods detects well-separated clusters, but does not solve the typical problem of the graph-based clustering algorithms (chaining affect). To dissolve this discrepancy, the fuzzy hyper volume measure is applied. If the cutting of the partition having the largest hyper volume cannot be executed based on Criterion-1 or Criterion-2, then the cut is performed based on the measure of the total fuzzy hyper volume. If this partition has N objects, then $N - 1$ possible cuts must be checked. Each of the $N - 1$ possibilities results in a binary split, hereby the objects placed in the cluster with the largest hyper volume are distributed into two subclusters. The algorithm chooses the binary split that results in the least total fuzzy hyper volume. The whole process is carried out until a termination criterion is satisfied (e.g., the predefined number of clusters, and/or the minimal number of objects in each partition is reached). As the number of the clusters is not known beforehand, it is suggested to give a relatively large threshold for it and then to draw the single linkage based dendrogram of the clusters to determine the proper number of them.

The application of this hybrid cutting criterion can be seen as a divisive hierarchical method. Following a depth-first tree-growing process, cuttings are iteratively performed. The final outcome is a hierarchical clustering tree, where the termination nodes are the final clusters. Figure 2.2 demonstrates a possible result after applying the different cutting methods on the MST. The partitions marked by the solid lines are resulted by the applying of the classical MST-based clustering methods (Criterion-1 or Criterion-2), and the partitions having gray dotted notations are arising from the application of the fuzzy hyper volume criterion (Criterion-3).

When compact parametric representation of the clusters is needed a Gaussian mixture model-based clustering should be performed where the number of Gaussians is equal to the termination nodes, and iterative Gath-Geva algorithm is initialized based on the partition obtained from the cuted MST. This approach is really fruitful, since it is well-known that the Gath-Geva algorithm is sensitive to the initialization of the partitions. The previously obtained clusters give an appropriate starting-point for the GG algorithm. Hereby, the iterative application of the Gath-Geva algorithm

Fig. 2.2 Binary tree given by the proposed Hybrid MST-GG algorithm

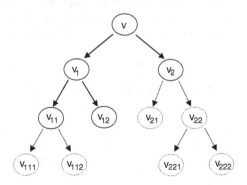

results in a good and compact representation of the clusters. The whole Hybrid MST-GG algorithm is described in Algorithm 7.

Algorithm 7 Hybrid MST-GG clustering algorithm

Step 0   Normalize the variables.

Step 1   Create the minimal spanning tree of the normalized objects.

Repeat   Iteration

    Step 2   Node selection. Select the node (i.e., subcluster) with the largest hyper volume V_i from the so-far formed hierarchical tree. Perform a cutting on this node based on the following criteria.

    Step 3   Binary Splitting.

- If the selected subcluster can be cut by Criterion-1, eliminate the edge with the largest weight that meets Criterion-1.
- If the selected subcluster cannot be cut by Criterion-1, but there exists an edge which corresponds to Criterion-2 perform a split. Eliminate the edge with the largest weight that meets Criterion-2.
- If the cluster having the largest hyper volume cannot be cut by Criterion-1 or Criterion-2, perform a split based on the following: Each of the edges in the corresponding subcluster with the largest volume in the so-far formed hierarchical tree is cut. With each cut, a binary split of the objects is formed. If the current node includes N_i objects, then $N_i - 1$ such splits are formed. The two subclusters, formed by the binary splitting, plus the clusters formed so far (excluding the current node) compose the potential partition. The total fuzzy hyper volume (FHV) of all formed $N_i - 1$ potential partitions are computed. The one that exhibits the lowest FHV is selected as the best partition of the objects in the current node.

Until   the termination criterion is satisfied (e.g. minimum number of objects in each subcluster and/or maximum number of clusters).

Step 4   When the compact parametric representation of the result of the clustering is needed, then Gath-Geva clustering is performed, where the number of the Gaussians is equal to the termination nodes, and the GG algorithm is initialized based on the partition obtained at the previous step.

The Hybrid MST-GG clustering method has the following four parameters: (i) cutting condition for the classical splitting of the MST (Criterion-1 and Criterion-2); (ii) terminating criterion for stopping the iterative cutting process; (iii) weighting exponent m of the fuzzy membership values (see GG algorithm in Appendix A.5), and (iv) termination tolerance ε of the GG algorithm.

2.2.2 Analysis and Application Examples

The previously introduced Hybrid MST-GG algorithm involves two major parts: (1) creating a clustering result based on the cluster volume based splitting extension of the basic MST-based clustering algorithm, and (2) utilizing this clustering output as initialization parameters in Gath-Geva clustering method. This way, the combined application of these major parts creates a fuzzy clustering.

The first part of the Hybrid MST-GG algorithm involves iterative cuttings of MST. The termination criterion of this iterative process can be based on the determination of the maximum number of clusters (c_{max}). When the number of the clusters is not known beforehand, it is suggested to determine this parameter a little larger than the expectations. Hereby, the Hybrid MST-GG algorithm would result in c_{max} fuzzy clusters. To determine the proper number of clusters it is worth drawing a dendrogram of the resulted clusters based on their similarities (e.g., single linkage, average linkage). Using these diagrams, the human 'data miner' can get a conception how similar the clusters are in the original space and is able to determine which clusters should be merged if it is needed. Finally, the resulted fuzzy clusters can also be converted into hard clustering result based on the fuzzy partition matrix by assigning the objects to the cluster characterized by the largest fuzzy membership value.

In the following clustering of some tailored data and well-known data sets are presented. In the following if not defined differently, parameters of GG method were chosen to be $m = 2$ and $\varepsilon = 0.0001$ according to the practice.

2.2.2.1 Handling the Chaining Effect

The first example is intended to illustrate that the proposed cluster volume based splitting extension of the basic MST-based clustering algorithm is able to handle (avoid) the chaining phenomena of the classical single linkage scheme. Figure 2.3 presents the minimal spanning tree of the normalized ChainLink data set (see Appendix A.6.9) and the result of the classical MST based clustering method. The value of parameter λ in this example was chosen to be 2. It means that based on Criterion-1 and Criterion-2, those edges are removed from the MST that are 2 times longer than the average length of the edges of the MST or 2 times longer than the average length of nearby (connected) edges. Parameter settings $\lambda = 2 \ldots 57$ give the same results. As

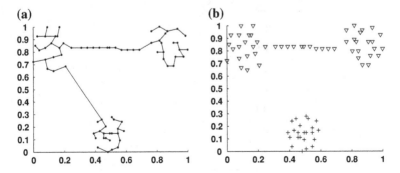

Fig. 2.3 Classical MST based clustering of ChainLink data set. **a** MST of the ChainLink data set. **b** Clusters obtained by the classical MST based clustering algorithm

Fig. 2.3b illustrates, the classical MST based algorithm detects only two clusters. If parameter λ is set to a smaller value, the algorithm cuts up the spherical clusters into more subclusters, but it does not unfold the chain link. If parameter λ is very large ($\lambda = 58, 59, \ldots$), the classical MST-based algorithm cannot separate the data set.

Figure 2.4 shows the results of the Hybrid MST-GG algorithm running on the normalized ChainLink data set. Parameters were set as follows: $c_{max} = 4$, $\lambda = 2$, $m = 2$, $\varepsilon = 0.0001$. Figure 2.4a shows the fuzzy sets that are the results of the Hybrid MST-GG algorithm. In this figure, the dots represent the data points and the 'o' markers are the cluster centers. The membership values are also shown, since the curves represent the isosurfaces of the membership values that are inversely proportional to the distances. It can be seen that the Hybrid MST-GG algorithm partitions the data set

Fig. 2.4 Result of the MST-GG clustering algorithm based on the ChainLink data set. **a** Result of the MST-GG clustering algorithm. **b** Hard clustering result obtained by the Hybrid MST-GG clustering algorithm

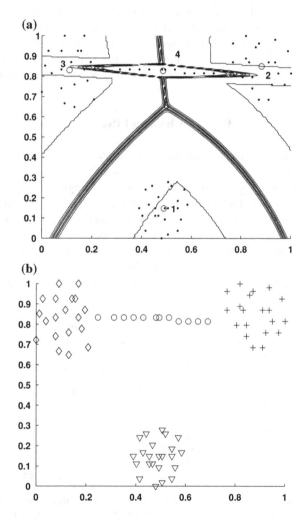

adequately, and it also unfolds the data chain between the clusters. Figure 2.4b shows the hard clustering result of the Hybrid MST-GG algorithm. Objects belonging to different clusters are marked with different notations. It is obtained by assigning the objects to the cluster characterized by the largest fuzzy membership value. It can be seen that the clustering rate is 100 %.

This short example illustrates the main benefit of the incorporation of the cluster validity based criterion into the classical MST based clustering algorithm. In the following, it will be shown how the resulting nonparametric clusters can be approximated by a mixture of Gaussians, and how this approach is beneficial for the initialization of these iterative partitional algorithms.

2.2.2.2 Handling the Convex Shapes of Clusters: Effect of the Initialization

Let us consider a more complex clustering problem with clusters of convex shape. This example is based on the Curves data set (see Appendix A.6.10). For the analysis, the maximum number of the clusters was chosen to be $c_{max} = 10$, and parameter λ was set to $\lambda = 2.5$. As Fig. 2.5 shows, the cutting of the MST based on the hybrid cutting criterion is able to detect properly clusters, because there is no partition containing data points from different curves. The partitioning of the clusters has not been stopped at the detection of the well-separated clusters (Criterion-1 and Criterion-2), but the resulting clusters have been further split to get clusters with small volumes, (Criterion-3). The main benefit of the resulted partitioning is that it can be easily approximated by a mixture of multivariate Gaussians (ellipsoids). This approximation is useful since the obtained Gaussians give a compact and parametric description of the clusters.

Figure 2.6a shows the final result of the Hybrid MST-GG clustering. The notation of this figure are the same as in Fig. 2.4. As can be seen, the clusters provide an

Fig. 2.5 Clusters obtained by cutting of the MST based on the hybrid cutting criterion (Curves data set)

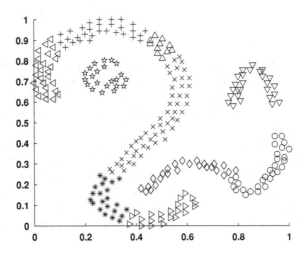

Fig. 2.6 Result of the Hybrid
MST-GG clustering algorithm
based on the Curves data set.
a Result of the Hybrid MST-
GG clustering algorithm.
b Single linkage dendrogram
based on the result of the
Hybrid MST-GG method

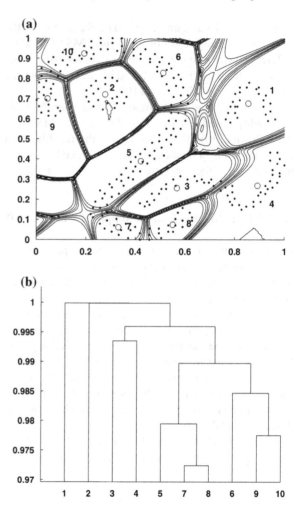

excellent description of the distribution of the data. The clusters with complex shape
are approximated by a set of ellipsoids. It is interesting to note, that this clustering
step only slightly modifies the placement of the clusters (see Figs. 2.5 and 2.6a).
To determine the adequate number of the clusters, the single linkage dendrogram
has been also drawn based on the similarities of the clusters. Figure 2.6b shows
that it is worth merging clusters '7' and '8', then clusters '9' and '10', following
this the merging of clusters {7, 8} and 5 is suggested, then follows the merging of
clusters {6} and {9, 10}. After this merging, the clusters {5, 7, 8} and {6, 9, 10} are
merged, hereby all objects placed in the long curve belongs to a single cluster. The
merging process can be continued based on the dendrogram. Halting this iterative
process at the similarity level 0.995, the resulted clusters meet the users' expectations
(clustering rate is 100 %).

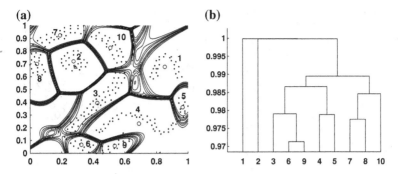

Fig. 2.7 Result of the Gath-Geva clustering initialized by fuzzy c-means (Curves data set). **a** Result of the GG clustering initialized by FCM. **b** Dendrogram based on the result of FCM-GG

For testing the effect of the parameters, we have performed several runs with different values of parameters λ and c_{max}.[1] It is not advisable to select parameter λ to be smaller than 2, because the data set is then cut up into many small subclusters. While choosing parameter λ to be greater than 2 does not have an effect on the final result. If c_{max} is chosen to be smaller than 10, the algorithm is not able to cut up the large ('S') curve. If parameter c_{max} is chosen to be larger than 10, the Hybrid MST-GG algorithm discloses the structure of the data set well.

In order to demonstrate the effectiveness of the proposed initialization scheme, Fig. 2.7 illustrates the result of the Gath-Geva clustering, where the clustering was initialized by the classical fuzzy c-means algorithm. As can be seen, this widely applied approach failed to find the proper clustering of the data set, only a sub-optimal solution has been found. The main difference between these two approaches can be seen in the dendrograms (see Figs. 2.6b and 2.7b).

2.2.2.3 Application to Real Life Data: Classification of Iris Flowers

The previous example showed that it is possible to obtain a properly clustered representation by the proposed mapping algorithm. However, the real advantage of the algorithm was not shown. This will be done by the clustering of the well-known Iris data set (see Appendix A.6.1). The parameters were set as follows: $c_{max} = 3$, $\lambda = 2.5$, $m = 2$ and $\varepsilon = 0.0001$.

The basic MST based clustering method (Criterion-1 and Criterion-2) detects only two clusters. In this case, the third cluster is formed only after the application of the cluster volume based splitting criterion (Criterion-3). The resulted three clusters correspond to the three classes of the Iris flowers. At the analysis of the distribution of the classes in the clusters, we found only three misclassification errors. The mixture of Gaussians density model is able to approximate this cluster arrangement. The

---

[1] The effect of parameters m and ε was not tested, because these parameters has effects only on the GG algorithm. These parameters were chosen as it suggested in [13].

Fig. 2.8 Converted result of the Hybrid MST-GG clustering on Iris data set

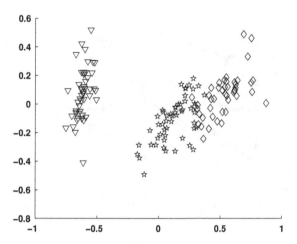

fuzzy clusters resulted by the Hybrid MST-GG algorithm were converted to a hard clustering by assigning each pattern to the cluster with the largest measure of membership. After this fine-tuning clustering step, we found only five misclassifications. This means 96.67 % classification correctness, that is a quite good result for this classification problem. Figure 2.8 shows the two-dimensional mapped visualization of the classified Iris data set based on the Hybrid MST-GG algorithm completed with the fuzzy-hard conversion. The two-dimensional mapping was made by the classical multidimensional scaling.

2.3 Jarvis-Patrick Clustering

While most similarity measures are based on distances defied in the n-dimensional vector space (e.g. Manhattan distances, Mahalanobis distance), similarity measures useful for topology-based clustering utilize neighborhood relations (e.g., mutual neighbor distance).

Jarvis-Patrick clustering (JP) [19] is a very simple clustering method. The algorithm first finds k nearest neighbors (knn) of all the objects. Two objects are placed in the same cluster whenever they fulfill the following two conditions:

- they must be each other's k-nearest neighbors, and
- they must have at least l nearest neighbors in common.

The algorithm has two parameters:

- parameter k, that is the number of the nearest neighbors to be taken into consideration, and
- parameter l, that determines the number of common neighbors necessary to classify two objects into the same cluster.

The main drawback of this algorithm is that the determination of the parameters k and l influences the output of the algorithm significantly. Other drawbacks are:

- the decision criterion is very rigid (the value of l), and
- this decision is constrained by the local k-nearest neighbors.

To avoid these disadvantages we suggested an extension of the similarity measure of the Jarvis-Patrick algorithm. The suggested *fuzzy neighborhood similarity measure* takes not only the k nearest neighbors into account, and it gives a nice tool to tune parameter l based on visualization and hierarchical clustering methods that utilize the proposed fuzzy neighborhood similarity. The proposed extension is carried out in the following two ways:

- fuzzyfication of parameter l, and
- spreading of the scope of parameter k.

The suggested fuzzy neighborhood similarity measure can be applied in various forms, in different clustering and visualization techniques (e.g. hierarchical clustering, MDS, VAT). In this chapter, some application examples are also introduced to illustrate the efficiency of the use of the proposed fuzzy neighborhood similarity measure in clustering. These examples show that the fuzzy neighborhood similarity measure based clustering techniques are able to detect clusters with different sizes, shapes, and densities. It is also shown that outliers are also detectable by the proposed measure.

2.3.1 Fuzzy Similarity Measures

Let $\mathbf{X} = \{\mathbf{x}_1, \mathbf{x}_2, \ldots, \mathbf{x}_N\}$ be the set of data. Denote $\mathbf{x}_i$ the ith object, which consists of D measured variables, grouped into an D-dimensional column vector $\mathbf{x}_i = [x_{1,i}, x_{2,i}, \ldots, x_{D,i}]^T$, $\mathbf{x}_i \in \mathbb{R}^D$. Denote $m_{i,j}$ the number of common k-nearest neighbors of $\mathbf{x}_i$ and $\mathbf{x}_j$. Furthermore, denote set A_i the k-nearest neighbors of $\mathbf{x}_i$, and A_j, respectively, for $\mathbf{x}_j$. The Jarvis-Patrick clustering groups $\mathbf{x}_i$ and $\mathbf{x}_j$ in the same cluster, if Eq. (2.9) holds.

$$\mathbf{x}_i \in A_j \quad \text{and} \quad \mathbf{x}_j \in A_i \quad \text{and} \quad m_{i,j} > l \tag{2.9}$$

Because $m_{i,j}$ can be expressed as $|A_i \cap A_j|$, where $|\bullet|$ denotes the cardinality, the $m_{i,j} > l$ formula is equivalent with the expression $|A_i \cap A_j| > l$. To refine decision criterion, a new similarity measure between the objects is suggested [28]. The proposed *fuzzy neighborhood similarity measure* is calculated in the following way:

$$s_{i,j} = \frac{|A_i \cap A_j|}{|A_i \cup A_j|} = \frac{m_{i,j}}{2k - m_{i,j}} \tag{2.10}$$

Equation (2.10) means that fuzzy neighborhood similarity characterizes similarity of a pair of objects by the fraction of the number of the common neighbors and the number of the total neighbors of that pair. The fuzzy neighborhood similarity measure is calculated between all pairs of objects, and it takes a value from [0, 1]. The $s_{i,j} = 1$ value indicates the strongest similarity between the objects, and the $s_{i,j} = 0$ expresses that objects x_i and x_j are very different from each other. Naturally, $s_{i,i} = 1$ for all $i = 1, 2, \ldots, N$. With the usage of the fuzzy neighborhood similarity measure the crisp parameter l of the Jarvis-Patrick algorithm is fuzzyfied.

Topology representation can be improved by more sophisticated neighborhood representation. The calculation of the *transitive neighborhood fuzzy similarity measure* is an iterative process. In each iteration step, there is an r-order similarity measure $(s_{i,j}^{(r)})$ calculated of the objects. In the case of $r = 1$ the $s_{i,j}^{(1)}$ is calculated as the fraction of the number of shared neighbors of the k-nearest neighbors of objects x_i and x_j and the total number of the k-nearest neighbors of objects x_i and x_j. In this case, Eq. (2.10) is obtained. Generally, $s_{i,j}^{(r)}$ is calculated in the following way:

$$s_{i,j}^{(r)} = \frac{|A_i^{(r)} \cap A_j^{(r)}|}{|A_i^{(r)} \cup A_j^{(r)}|}, \tag{2.11}$$

where set $A_i^{(r)}$ denotes the r-order k-nearest neighbors of object x_i, and $A_j^{(r)}$, respectively, for x_j. In each iteration step, the pairwise calculated fuzzy neighborhood similarity measures are updated based on the following formula:

$$s_{i,j}^{\cdot(r)} = (1 - \alpha)s_{i,j}^{\cdot(r-1)} + \alpha s_{i,j}^{(r)}, \tag{2.12}$$

where α is the first-order filter parameter. The iteration process proceeds until r reaches the predefined value (r_{max}). The whole procedure is given in Algorithm 8.

As a result of the whole process, a *fuzzy neighborhood similarity matrix* (**S**) will be given containing pairwise fuzzy neighborhood similarities. The *fuzzy neighborhood distance matrix* (**D**) of the objects is obtained by the formula: $\mathbf{D} = 1 - \mathbf{S}$. These similarity distance matrices are symmetrical, $\mathbf{S}^T = \mathbf{S}$ and $\mathbf{D}^T = \mathbf{D}$.

The computation of the proposed transitive fuzzy neighborhood similarity/distance measure includes the proper setting of three parameters: k, r_{max}, and α. Lower k (e.g., $k = 3$) separate the clusters better. By increasing value k clusters will overlap in similar objects. The higher r_{max} is, the higher the similarity measure becomes. Increase of r_{max} results in more compact clusters. The lower the value of α, the less the affect of far neighbors becomes.

As the fuzzy neighborhood similarity measure is a special case of the transitive fuzzy neighborhood similarity measure in the following these terms will be used as equivalent.

Algorithm 8 Calculation of the transitive fuzzy neighborhood similarity measure

Given a set of data $\mathbf{X}$, specify the number of the maximum clusters r_{max}, and choose a first-order filter parameter α. Initialize the fuzzy neighborhood similarity matrix as $\mathbf{S}^{(0)} = 0$.

Repeat for $r = 1, 2, \ldots, r_{max}$

Step 1 Calculate the fuzzy neighborhood similarities for each pair of objects as follows:

$$s_{i,j}^{(r)} = \frac{|A_i^{(r)} \cap A_j^{(r)}|}{|A_i^{(r)} \cup A_j^{(r)}|}, \tag{2.13}$$

where set $A_i^{(r)}$ denotes the r-order k-nearest neighbors of object $\mathbf{x}_i \in \mathbf{X}$, and $A_j^{(r)}$, respectively, for $\mathbf{x}_j \in \mathbf{X}$.

Step 2 Update the fuzzy neighborhood similarity measures based on the following formula:

$$s_{i,j}^{'(r)} = (1 - \alpha)s_{i,j}^{'(r-1)} + \alpha s_{i,j}^{(r)}, \tag{2.14}$$

Finally, $s_{i,j}^{'(r_{max})}$ yields the fuzzy neighborhood similarities of the objects.

2.3.2 Application of Fuzzy Similarity Measures

There are several ways to apply the previously introduced fuzzy neighborhood similarity/distance matrix. For example, hierarchical clustering methods work on similarity or distance matrices. Generally, these matrices are obtained from the Euclidian distances of pairs of objects. Instead of the other similarity/distance matrices, the hierarchical methods can also utilize the fuzzy neighborhood similarity/distance matrix. The dendrogram not only shows the whole iteration process, but it can also be a useful tool to determine the number of the data groups and the threshold of the separation of the clusters. To separate the clusters, we suggest to draw the fuzzy neighborhood similarity based dendrogram of the data, where the long nodes denote the proper threshold to separate the clusters.

The visualization of the objects may significantly assist in revealing the clusters. Many visualization techniques are based on the pairwise distance of the data. Because multidimensional scaling methods (MDS) (see Sect. 3.3.3) work on dissimilarity matrices, this method can also be based on the fuzzy neighborhood distance matrix. Furthermore, the VAT is also an effective tool to determine the number of the clusters. Because VAT works with the dissimilarities of the data, it can also be based on the fuzzy neighborhood distance matrix.

In the following section, some examples are presented to show the application of the fuzzy neighborhood similarity/distance measure. The first example is based on a synthetic data set, and the second and third examples deal with visualization and clustering of the well-known Iris and Wine data sets.

The variety data set is a synthetic data set which contains 100 two-dimensional data objects. 99 objects are partitioned in 3 clusters with different sizes (22, 26, and 51 objects), shapes, and densities, and it also contains an outlier (see Appendix A.6.8).

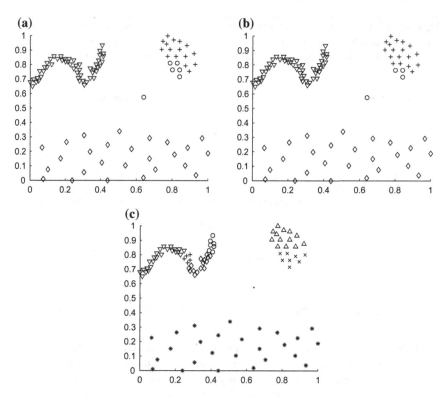

Fig. 2.9 Results of the Jarvis-Patrick clustering on the normalized Variety data set. **a** $k = 8, l = 3$. **b** $k = 8, l = 4$. **c** $k = 8, l = 5$

Figure 2.9 shows some results of Jarvis-Patrick clustering applied on the normalized data set. The objects belonging to different clusters are marked with different markers. In these cases, the value of parameter k was fixed to 8, and the value of parameter l was changed from 2 to 5. (The parameter settings $k = 8, l = 2$ gives the same result as $k = 8$ and $l = 3$.) It can be seen that the Jarvis-Patrick algorithm was not able to identify the clusters in any of the cases. The cluster placed in the upper right corner in all cases is split into subclusters. When parameter l is low ($l = 2, 3, 4$), the algorithm is not able to detect the outlier. When parameter l is higher, the algorithm detects the outlier, but the other clusters are split into more subclusters. After multiple runs of the JP algorithm, there appeared a clustering result, where all objects were clustered according to expectations. This parameter setting was: $k = 10$ and $l = 5$. To show the complexity of this data set in Fig. 2.10, the result of the well-known k-means clustering is also presented (the number of the clusters is 4). This algorithm is not able to disclose the outlier, thereby the cluster with small density is split into two subclusters.

Fig. 2.10 Result of the k-means clustering on the normalized Variety data set

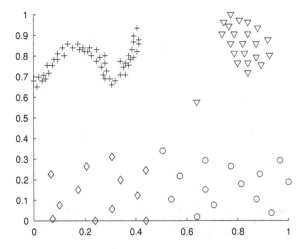

Table 2.1 Clustering rates for different mappings of the Variety data set

Algorithm	Clustering rate (%)
JP $k = 8\ l = 3$	95
JP $k = 8\ l = 4$	98
JP $k = 8\ l = 5$	65
JP $k = 10\ l = 5$	100
k-means	88

Table 2.1 summarizes the clustering rates of the previously presented algorithms. The clustering rate was calculated as the fraction of the number of well-clustered objects and the total number of objects.

The proposed fuzzy neighborhood similarity measure was calculated with different k, r_{max} and α parameters. Different runs with parameters $k = 3 \ldots 25$, $r_{max} = 2 \ldots 5$ and $\alpha = 0.1 \ldots 0.4$ have been resulted in good clustering outcomes. If a large value is chosen for parameter k, it is necessary to keep parameter r_{max} on a small value to avoid merging the outlier object with one of the clusters.

To show the fuzzy neighborhood distances of the data, the objects are visualized by multidimensional scaling and VAT. Figure 2.11a shows the MDS mapping of the fuzzy neighborhood distances with the parameter settings: $k = 6$, $r_{max} = 3$ and $\alpha = 0.2$. Other parameter settings have also been tried, and they show similar results to Fig. 2.11a. It can be seen that the calculated pairwise fuzzy neighborhood similarity measure separates the three clusters and the outlier well. Figure 2.11b shows the VAT representation of the data set based on the single linkage fuzzy neighborhood distances. The three clusters and the outlier are also easily separable in this figure.

To find the proper similarity threshold to separate the clusters and the outlier, the dendrogram based on the single linkage connections of the fuzzy neighborhood distances of the objects (Fig. 2.12) has also been drawn. The dendrogram shows that the value $d_{i,j} = 0.75$ $(d_{i,j} = 1 - s_{i,j})$ is a suitable choice to separate the clusters and

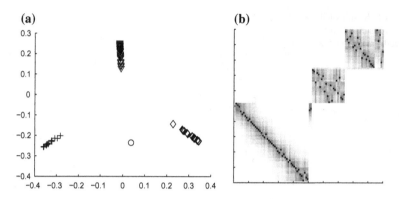

Fig. 2.11 Different graphical representations of the fuzzy neighborhood distances (Variety data set). **a** MDS based on the fuzzy neighborhood distance matrix. **b** VAT based on the single linkage fuzzy neighborhood distances

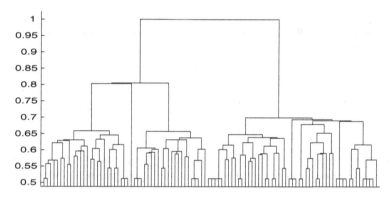

Fig. 2.12 Single linkage dendrogram based on the fuzzy neighborhood distances (Variety data set)

the outlier from each other ($k = 6$, $r_{max} = 3$ and $\alpha = 0.2$). Applying a single linkage agglomerative hierarchical algorithm based on the fuzzy neighborhood distances, and halting this algorithm at the threshold $d_{i,j} = 0.75$ the clustering rate is 100 %. In other cases ($k = 3 \ldots 25$, $r_{max} = 2 \ldots 5$ and $\alpha = 0.1 \ldots 0.4$, and if the value of parameter k was large, the parameter r_{max} was kept on low values), the clusters also were easily separable and the clustering rate obtained was 99–100 %.

This simple example illustrates that the proposed fuzzy neighborhood similarity measure is able to separate clusters with different sizes, shapes, and densities, furthermore it is able to identify outliers. The Wine database (see Appendix A.6.3) consists of the chemical analysis of 178 wines from 3 different cultivars in the same Italian region. Each wine is characterized by 13 attributes, and there are 3 classes distinguished. Figure 2.13 shows the MDS projections based on the Euclidian and the fuzzy neighborhood distances ($k = 6$, $r_{max} = 3$, $\alpha = 0.2$). The figures illustrate that the fuzzy neighborhood distance based MDS separates the three clusters better.

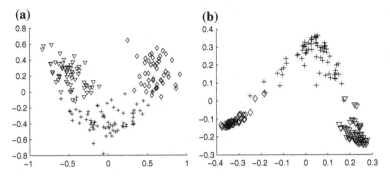

Fig. 2.13 Different MDS representations of the Wine data set. **a** MDS based on the Euclidian distances. **b** MDS based on the fuzzy neighborhood distance matrix

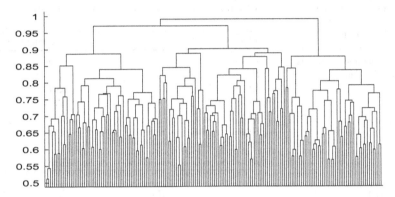

Fig. 2.14 Average linkage based dendrogram of fuzzy neighborhood distances (Wine data set)

To separate the clusters, we have drawn dendrograms based on the single, average, and the complete linkage distances. Using these parameters, the best result (clustering rate 96.62%) is given by the average linkage based dendrogram, on which the clusters are uniquely separable. In Fig. 2.14, the average linkage based dendrogram of the fuzzy neighborhood distances is shown. Figure 2.15 shows the VAT representation of the Wine data set based on the average linkage based relations of the fuzzy neighborhood distances. It can be see that the VAT representation also suggest to draw three clusters.

For the comparison the Jarvis-Patrick algorithm was also tested with different settings on this data set. Running results of this algorithm show very diverse clustering rates (see Table 2.2). The fuzzy neighborhood similarity was also tested on the Iris data set. This data set contains data about three types of iris flowers (see Appendix A.6.1). Iris setosa is easily distinguishable from the other two types, but the Iris versicolor and the Iris virginica are very similar to each other. Figure 2.16a shows the MDS mapping of this data set based on the fuzzy neighborhood distances. This visualization distinguishes the Iris setosa from the other two types, but the individ-

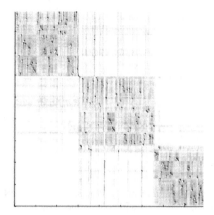

Fig. 2.15 Average linkage based VAT of fuzzy neighborhood distances (Wine data set)

Table 2.2 Clustering rates of Jarvis-Patrick algorithm at different parameter settings on the Wine data set

Algorithm	Clustering rate (%)
JP $k = 10\ l = 5$	30.34
JP $k = 10\ l = 4$	67.98
JP $k = 10\ l = 3$	76.97
JP $k = 10\ l = 2$	72.47
JP $k = 8\ l = 5$	16.29
JP $k = 8\ l = 4$	29.78
JP $k = 8\ l = 3$	69.67
JP $k = 8\ l = 2$	65.17

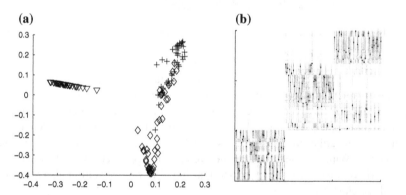

(a)　　　　　　　　　　　**(b)**

Fig. 2.16 MDS and VAT representations of the Iris data set based on the fuzzy neighborhood distances. **a** MDS based on the fuzzy neighborhood distance matrix. **b** VAT based on the fuzzy neighborhood distance matrix

uals of the Iris versicolor and virginica overlap each other. Figure 2.16b shows the VAT visualization of the fuzzy neighborhood distances based on the single linkage relations of the objects. The VAT visualization also suggests a well-separated and two overlapping clusters. The parameter settings in both cases were: $k = 5$, $r_{\max} = 3$

and $\alpha = 0.2$. Different runs of the original Jarvis-Patrick clustering have not given an acceptable result.

2.4 Summary of Graph-Based Clustering Algorithms

In this chapter, two graph-based clustering approaches were presented.

- The best-known graph-theoretic divisive clustering algorithm deletes the longest edges of minimal spanning tree (MST) of the data [30]. We introduced a new splitting criterion to improve the performance of clustering methods based on this principle. The proposed splitting criterion is based on the calculation of the hypervolume of the clusters. From the suggested splitting criterion follows that the resulted clusters can be easily approximated by multivariate Gaussian functions. Result of the cutting of the MST based on the combined cutting criteria (classical cutting and the proposed cutting) can be effectively used for the initialization of Gaussian mixture model-based clustering algorithms. The resulted Hybrid MST-GG clustering algorithm combines the graph-theoretic and the partitional model based clustering. The approach is demonstrated through some sets of tailored data and through the well-known Iris benchmark classification problem. The results showed the advantages of the hybridization of the hierarchial graph-theoretic and partitional model based clustering algorithm. It has been shown that: (1) the chaining problem of the classical MST based clustering has been solved; (2) the initialization of the Gath-Geva clustering algorithms has been properly handled, and (3) the resulting clusters are easily interpretable with the compact parametric description of the multivariate Gaussian clusters (fuzzy covariance matrices).
- The similarity of objects can be calculated based on neighborhood relations. The fuzzy neighborhood similarity measure introduced in this chapter extends the similarity measure of the Jarvis-Patrick algorithm in two ways: (i) it takes into account the far neighbors partway and (ii) it fuzzifies the crisp decision criterion of the Jarvis-Patrick algorithm.

The fuzzy neighborhood similarity measure is based on the common neighbors of the objects, but differently from the Jarvis-Patrick algorithm it is not restricted to the direct neighbors. While the fuzzy neighborhood similarity measure describes the similarities of the objects, the fuzzy neighborhood distance measure characterizes the dissimilarities of the data. The values of the fuzzy neighborhood distances are easily computable from the fuzzy neighborhood similarities. The application possibilities of the fuzzy neighborhood similarity and distance measures are widespread. All methods that work on distance or similarity measures can also be based on the fuzzy neighborhood similarity/distance measures. This chapter introduced the application possibilities of the fuzzy neighborhood similarity and distance measures in hierarchical clustering and in VAT representation. It was demonstrated through application examples that clustering methods based on the fuzzy neighborhood similarity/distance measure can discover clusters with arbitrary shapes, sizes, and densities. Furthermore, the fuzzy neighborhood similarity/distance measure is able to identify outliers, as well.

References

1. Anders, K.H.: A hierarchical graph-clustering approach to find groups of objects. In: Proceedings 5'th ICA workshop on progress in automated map generalization, IGN, pp. 28–30 (2003)
2. Augustson, J.G., Minker, J.: An analysis of some graph theoretical clustering techniques. J. ACM **17**, 571–588 (1970)
3. Backer, F.B., Hubert, L.J.: A graph-theoretic approach to goodness-of-fit in complete-link hierarchical clustering. J. Am. Stat. Assoc. **71**, 870–878 (1976)
4. Barrow, J.D., Bhavsar, S.P., Sonoda, D.H.: Minimal spanning trees, filaments and galaxy clustering. Mon. Not. R. Astron. Soc. **216**, 17–35 (1985)
5. Bezdek, J.C., Clarke, L.P., Silbiger, M.L., Arrington, J.A., Bensaid, A.M., Hall, L.O., Murtagh, R.F.: Validity-guided (re)clustering with applications to image segmentation. IEEE Trans. Fuzzy Syst. **4**, 112–123 (1996)
6. Bezdek, J., Pal, N.: Some new indexes of cluster validity. IEEE Trans. Syst. Man Cybern. **28**, 301–315 (1998)
7. Doman, T.N., Cibulskis, J.M., Cibulskis, M.J., McCray, P.D., Spangler, D.P.: Algorithm5: a technique for fuzzy similarity clustering of chemical inventories. J. Chem. Inf. Comput. Sci. **36**, 1195–1204 (1996)
8. Dunn, C.: Well separated clusters and optimal fuzzy partitions. J. Cybern. **4**, 95–104 (1974)
9. Eppstein, D., Paterson, M.S., Yao, F.F.: On nearest-neighbor graphs. Discrete Comput. Geom. **17**, 263–282 (1997)
10. Forina, M., Oliveros, C., Concepción, M., Casolino, C., Casale, M.: Minimum spanning tree: ordering edges to identify clustering structure. Anal. Chim. Acta **515**, 43–53 (2004)
11. Fortune, S.: Voronoi diagrams and delaunay triangulations. In: Du, D.-Z., Hwang, F.K. (eds.), Computing in Euclidean Geometry, pp. 193–223. World Scientific, Singapore (1992)
12. Gabriel, K., Sokal, R.: A new statistical approach to geographic variation analysis. Syst. Zool. **18**, 259–278 (1969)
13. Gath, I., Geva, A.B.: Unsupervised optimal fuzzy clustering. IEEE Trans. Pattern Anal. Mach. Intell. **11**, 773–781 (1989)
14. Gonzáles-Barrios, J.M., Quiroz, A.J.: A clustering procedure based on the comparsion between the k nearest neighbors graph and the minimal spanning tree. Stat. Probab. Lett. **62**, 23–34 (2003)
15. Gower, J.C., Ross, G.J.S.: Minimal spanning trees and single linkage cluster analysis. Appl. Stat. **18**, 54–64 (1969)
16. Guha, S., Rastogi, R., Shim, K.: ROCK: a robust clustering algorithm for categorical attributes. In: Proceedings of the 15th international conference on data engeneering, pp. 512–521 (1999)
17. Huang, X., Lai, W.: Clustering graphs for visualization via node similarities. J. Vis. Lang. Comput. **17**, 225–253 (2006)
18. Jaromczyk, J.W., Toussaint, G.T.: Relative neighborhood graphs and their relatives. Proc. IEEE **80**(9), 1502–1517 (1992)
19. Jarvis, R.A., Patrick, E.A.: Clustering using a similarity measure based on shared near neighbors. IEEE Trans. Comput. **C22**, 1025–1034 (1973)
20. Karypis, G., Han, E.-H., Kumar, V.: Chameleon: hierarchical clustering using dynamic modeling. IEEE Comput. **32**(8), 68–75 (1999)
21. Kruskal, J.B.: On the shortest spanning subtree of a graph and the traveling salesman problem. Proc. Am. Math. Soc. **7**(1), 48–50 (1956)

22. Päivinen, N.: Clustering with a minimum spanning tree of scale-free-like structure. Pattern Recog. Lett. **26**, 921–930 (2005)
23. Prim, R.C.: Shortest connection networks and some generalizations. Bell Syst. Tech. J. **36**, 1389–1401 (1957)
24. Raghavan, V.V., Yu, C.T.: A comparison of the stability characteristics of some graph theoretic clustering methods. IEEE Trans. Pattern Anal. Mach. Intell. **3**, 393–402 (1980)
25. Toussaint, G.T.: The relative neighborhood graph of a finite planar set. Pattern Recogn. **12**, 261–268 (1980)
26. Varma, S., Simon, R.: Iterative class discovery and feature selection using Minimal Spanning Trees. BMC Bioinform. **5**, 126–134 (2004)
27. Vathy-Fogarassy, A., Kiss, A., Abonyi, J.: Hybrid minimal spanning tree and mixture of Gaussians based clustering algorithm. In: Lecture Notes in Computer Science: Foundations of Information and Knowledge Systems vol. 3861, pp. 313–330. Springer, Heidelberg (2006)
28. Vathy-Fogarassy, A., Kiss, A., Abonyi, J.: Improvement of Jarvis-Patrick clustering based on fuzzy similarity. In: Masulli, F., Mitra, S., Pasi, G. (eds.) Applications of Fuzzy Sets Theory, LNCS, vol. 4578, pp. 195–202. Springer, Heidelberg (2007)
29. Yao, A.: On constructing minimum spanning trees in k-dimensional spaces and related problems. SIAM J. Comput. **11**, 721–736 (1892)
30. Zahn, C.T.: Graph-theoretical methods for detecting and describing gestalt clusters. IEEE Trans. Comput. **C20**, 68–86 (1971)

Chapter 3
Graph-Based Visualisation of High Dimensional Data

Abstract In this chapter we give an overview of classical dimensionality reduction and graph based visualisation methods that are able to uncover hidden structure of high dimensional data and visualise it in a low-dimensional vector space.

3.1 Problem of Dimensionality Reduction

In practical data mining problems high-dimensional data has to be analysed. Objects to be analysed are characterised with D variables, and each variable corresponds to a dimension of the vector space. So data characterised with D variables can be interpreted as a point in the D-dimensional vector space. The question is how can we visualise high-dimensional ($D > 3$) data for human eyes?

Since humans are can able to handle more then three dimensions, it is useful to map the high-dimensional data points into a low-dimensional vector space. Since we are particularly good at detecting certain patterns in visualised form, dimensionality reduction methods play an important role in exploratory data analysis [1] and visual data mining.

Dimensional reduction methods visualise high-dimensional objects in a lower dimensionality vector space in such a way that they try to preserve their spatial position. Therefore the mapping procedure is influenced by the spatial position of the objects to be analysed. The spatial distribution of the data points can be determined by the distances and/or the neighbourhood relations of the objects. In this way computing the pairwise distances plays an important role in the low-dimensional visualisation process. Furthermore, we can see if the cardinality of the assessed data is really high, the calculation of the pairwise distances requires notable time and computational cost. In such a cases the application of a vector quantisation method may further improve the quality of the depiction.

Based on this principles the mapping-based visualisation algorithms can be defined as combinations of:

Á. Vathy-Fogarassy and J. Abonyi, *Graph-Based Clustering
and Data Visualization Algorithms*, SpringerBriefs in Computer Science,
DOI: 10.1007/978-1-4471-5158-6_3, © János Abonyi 2013

1. vector quantisation (optional step),
2. calculation of the similarities or dissimilarities among the quantised or original objects,
3. mapping of the quantised or original objects.

Each steps can be performed in a number of ways, and so dimensionality reduction based visualisation methods offer a wide repository of opportunities. In the following let take a closer look at applicable methods:

- *Vector quantisation* (VQ): There are several ways to reduce the number of the analysed objects. Most often the presented k-means [2] or neural gas [3] algorithms are used (see Chap. 1).
- *Calculation of similarities or dissimilarities*: Since applied mapping methods work on similarity/dissimilarty matrices of the objects the computation of the pairwise similarities or dissimilarities of the objects influences basically the result of the mapping process. Dissimilarities can be expressed as distances. Distances among the objects or quantised data can be calculated based on either a distance norm (e.g. Euclidean norm) or based on a graph of the data. Calculation of graph based distances requires the exploration of the topology of the objects or its representatives. For this purpose the graph of the original or quantised data points should be obtained by

  - the ε-neighbouring,
  - the k-neighbouring,
  - or can be given by other algorithms used to generate the topology of data points (for example, topology representing networks).

- *Mapping*: The mapping of the quantised or original data can also be carried out in different ways (e.g. multidimensional scaling, Sammon mapping, principal component analysis, etc.). These methods differ from each other in basic computation principles, and therefore they can results in fairly different visualisation results. Section 3.1 gives a summary from these methods.

As we previously mentioned, visualisation of high-dimensional data can be built up as a combination of data reduction and dimensionality reduction. The goal of *dimensionality reduction* is to map a set of observations from a high-dimensional space (D) into a low-dimensional space (d, $d \ll D$) preserving as much of the intrinsic structure of the data as possible. Let $\mathbf{X} = \{\mathbf{x}_1, \mathbf{x}_2, \ldots, \mathbf{x}_N\}$ be a set of the observed data, where $\mathbf{x}_i$ denotes the i-th observation ($\mathbf{x}_i = [x_{i,1}, x_{i,2}, \ldots, x_{i,D}]^T$). Each data object is characterised by D dimensions, so $x_{i,j}$ yields the j-th ($j = 1, 2, \ldots, D$) attribute of the i-th ($i = 1, 2, \ldots, N$) data object. Dimensionality reduction techniques transform data set $\mathbf{X}$ into a new data set $\mathbf{Y}$ with dimensionality d ($\mathbf{Y} = \{\mathbf{y}_1, \mathbf{y}_2, \ldots, \mathbf{y}_N\}$, $\mathbf{y}_i = [y_{i,1}, y_{i,2}, \ldots, y_{i,d}]^T$). In the reduced space many data analysis tasks (e.g. classification, clustering, image recognition) can be carried out faster than in the original data space.

Dimensionality reduction methods can be performed in two ways: they can apply *feature selection* algorithms or they can based on different *feature extraction* methods. The topic of feature extraction and features selection methods is an active research area, lots of research papers introduce new algorithms or utilise them in different scientific fields (e.g. [4–9]).

Feature selection methods keep most important dimensions of the data and eliminate unimportant or noisy factors. *Forward selection methods* start with an empty set and add variables to this set one by one by optimizing an error criterion. *Backward selection methods* start with all variables in the selected set and remove them one by one, in each step removing the one that decreases the error the most.

In the literature there are many approaches (e.g. [10–13]) described to select the proper subset of the attributes. The well known exhaustive search method [12] examines all possible $\binom{D}{d}$ subsets and selects the subset with largest feature selection criterion as the solution. This method guarantees to find the optimum solution, but if the number of the possible subsets is large, it becomes impractical. There have been many methods proposed to avoid the enormous computational cost (e.g. branch and bound search [14], floating search [15], Monte Carlo algorithms).

Feature extraction methods do not select the most relevant attributes but they combine them into some new attributes. The number of these new attributes is generally more less than the number of the original attributes. So feature extraction methods take all attributes into account and they provide reduced representation by feature combination and/or transformation. The resulted representation provides relevant information about the data. There are several dimensionality reduction methods proposed in the literature based on the feature extraction approach, for example the well known Principal Component Analysis (PCA) [16, 17], Sammon mapping (SM) [18], or the Isomap [19] algorithm.

Data sets to be analysed often contain lower dimensional manifolds embedded in higher dimensional space. If these manifolds are linearly embedded into high-dimensional vector space the classical *linear dimensionality reduction methods* provide a fairly good low-dimensional representation of data. These methods assume that data lie on a linear or on a near linear subspace of the high-dimensional space and they calculate the new coordinates of data as the linear combination of the original variables. The most commonly used linear dimensionality reduction methods are for example the Principal Component Analysis (PCA) [17], the Independent Component Analysis (ICA) [20] or the Linear Discriminant Analysis (LDA) [21]. However if the manifolds are nonlinearly embedded into the higher dimensional space linear methods provide unsatisfactory representation of data. In these cases the *nonlinear dimensionality reduction methods* may outperform the traditional linear techniques and they are able to give a good representation of data set in the low-dimensional data space. To unfold these nonlinearly embedded manifolds many nonlinear dimensionality reduction methods are based on the concept of geodesic distance and they build up graphs to carry out the visualisation process (e.g. Isomap, Isotop, TRNMap). The bets known nonlinear dimensionality reduction methods are Kohonen's Self-Organizing Maps (SOM) [22, 23], Sammon mapping [18], Locally Linear Embedding (LLE) [24, 25], Laplacian Eigenmaps [26] or Isomap [19].

Dimensional reduction methods approximate high-dimensional data distribution in a low-dimensional vector space. Different dimensionality reduction approaches emphasise different characteristics of data sets to be mapped. Some of them try to preserve pairwise distances of the objects, others try to preserve global ordering or neighbourhoud relations of data, and again others are based on mathematical models representing other functional relationships. Based on the applied principle of the mapping three types of the dimensionality reduction methods can be distinguished:

- *metric methods* try to preserve the distances of the data defined by a metric,
- *non-metric methods* try to preserve the global ordering relations of the data,
- *other methods* that are based on different mathematical approaches and differ from the previously introduced two groups.

As distance based metric methods may provide less good results in neighbourhood preservation. Naturally, non-metric dimensionality reduction methods emphasise the preservation of the neighbourhood relations, and therefore they provide lower performance in distance preservation. The quality of the mapping can be measured in different ways. In the next subsection these quality measures will be introduced.

3.2 Measures of the Mapping Quality

During the mapping process algorithms try to approximate high-dimensional spatial distribution of the objects in a low-dimensional vector space. Different algorithms may result in different low-dimensional visualisations by emphasising different characteristics of the objects relationships. While some of them try the preserve distances others neighbourhood relationships. From an other aspects we can see that there are algorithms that emphasise the local structure of the data points, while other methods put the global structure in the focus. According to these approaches the evaluation criteria of mapping can be summarised as follows:

- *Distance versus neighbourhood preservation*: Mapping methods try to preserve either the distances or the neighbourhood relations among the data points. While dimensionality reduction methods based on distance preservation try to preserve the pairwise distances between the samples, the mapping methods based on neighbourhood preservation attempt to preserve the global ordering relation of the data. There are several numeral measures proposed to express how well the distances are preserved, for example the classical metric MDS [27, 28] and the Sammon stress [18] functions. The degree of the preservation of the neighbourhood relations can be measured by the functions of trustworthiness and continuity [29, 30].
- *Local versus global methods*. On the other hand, the analysis of the considered mapping methods can be based on the evaluation of the mapping qualities in local and global environment of the objects. Local approaches attempt to preserve the local geometry of data, namely they try to map nearby points in the input space to nearby points in the output space. Global approaches attempt to preserve geometry at all scales by mapping nearby points in the input space to nearby points in the output space, and faraway points in the input space to faraway points in the output space.

To measure distance preservation of the mapping methods mostly the Sammon stress function, classical MDS stress function and residual variance are used most commonly.

Sammon stress and *classical MDS stress functions* are similar to each other. Both functions calculate pairwise distances in the original and in the reduced low-dimensional space as well. Both measures can be interpreted as an error between original distances in the high-dimensional vector space and the mapped low-dimensional vector space. The difference between the Sammon stress and the classical MDS stress functions is that the Sammon stress contains a normalizing factor. In the Sammon stress errors are normalised by distances of the input data objects. The Sammon stress is calculated as it is shown in Eqs. 3.1 and 3.2 demonstrate the classical MDS stress function.

$$E_{SM} = \frac{1}{\sum\limits_{i<j} d_{i,j}^*} \sum_{i<j}^{N} \frac{(d_{i,j}^* - d_{i,j})^2}{d_{i,j}^*}, \tag{3.1}$$

$$E_{metric\_MDS} = \frac{1}{\sum\limits_{i<j} d_{i,j}^{*2}} \sum_{i<j}^{N} (d_{i,j}^* - d_{i,j})^2, \tag{3.2}$$

In both equations $d_{i,j}^*$ denotes the distance between the ith and jth original objects, and $d_{i,j}$ yields the distance for the mapped data points in the reduced vector space. Variable N yields the number of the objects to be mapped.

The error measure is based on the *residual variance* defined as:

$$1 - R^2(D_X^*, D_Y), \tag{3.3}$$

where D_Y denotes the matrix of Euclidean distances in the low-dimensional output space ($D_Y = [d_{i,j}]$), and D_X^*, $D_X^* = [d_{i,j}^*]$ is the best estimation of the distances of the data to be projected. The pairwise dissimilarities of the objects in the input space may arising from the Euclidean distances or may be estimated by graph distances of the objects. R is the standard linear correlation coefficient, taken over all entries of D_X^* and D_Y.

In the following in this book, when the examined methods utilise geodesic or a graph distances to calculate the pairwise dissimilarities of the objects in the high-dimensional space the values of the dissimilarities of these objects ($d_{i,j}^*$) are also estimated based on this principle.

The neighbourhood preservation of the mappings and the local and global mapping qualities can be measured by functions of trustworthiness and continuity. Kaski and Vienna pointed out that every visualisation method has to make a tradeoff between gaining good trustworthiness and preserving the continuity of the mapping [30, 31].

A projection is said to be *trustworthy* [29, 30] when the nearest neighbours of a point in the reduced space are also close in the original vector space. Let N be the number of the objects to be mapped, $U_k(i)$ be the set of points that are in the k size neighbourhood of the sample i in the visualisation display but not in the original data space. Trustworthiness of visualisation can be calculated in the following way:

$$M_1(k) = 1 - \frac{2}{Nk(2N - 3k - 1)} \sum_{i=1}^{N} \sum_{j \in U_k(i)} (r(i, j) - k), \qquad (3.4)$$

where $r(i, j)$ denotes the ranking of the objects in input space.

The projection onto a lower dimensional output space is said to be *continuous* [29, 30] when points near to each other in the original space are also nearby in the output space. Continuity of visualisation is calculated by the following equation:

$$M_2(k) = 1 - \frac{2}{Nk(2N - 3k - 1)} \sum_{i=1}^{N} \sum_{j \in V_k(i)} (s(i, j) - k), \qquad (3.5)$$

where $s(i, j)$ is the rank of the data sample i from j in the output space, and $V_i(k)$ denotes the set of those data points that belong to the k-neighbours of data sample i in the original space, but not in the mapped space used for visualisation.

In this book when mappings are based on geodesic distances, the ranking values of the objects in both cases (trustworthiness and continuity) are calculated based on the geodesic distances.

Mapping quality of the applied methods in local and in global area can be expressed by trustworthiness and continuity. Both measures are function of the number of neighbours k. Usually, trustworthiness and continuity are calculated for $k = 1, 2, \ldots, k_{max}$, where k_{max} denotes the maximum number of the objects to be taken into account. At small values of parameter k the local reconstruction performance of the model can be tested, while at larger values of parameter k the global reconstruction is measured.

Topographic error and topographic product quality measures may also be used to give information about the neighbourhood preservation of mapping algorithms. *Topographic error* [32] takes only the first and second neighbours of each data point into account and it analyzes whether the nearest and the second nearest neighbours remain neighbours of the object in the mapped space or not. If these data points are not adjacent in the mapped graph the quality measure considers this a mapping error. The sum of errors is normalized to a range from 0 to 1, where 0 means the perfect topology preservation.

Topographic product introduced by Bauer in 1992 [33] was developed for qualifying the mapping result of SOM. This measure has an input parameter k and it takes not only the two nearest neighbor into account. The topographic product compares the neighbourhood relationship between each pair of data points with respect to both their position in the resulted map and their original reference vectors in the observation space. As result it indicates whether the dimensionality of the output space is too small or too large.

3.3 Standard Dimensionality Reduction Methods

Although standard dimensionality reduction methods do not utilise neighbourhood relations of data points and they do not build graphs to represent the topology, many graph based visualisation methods utilise these techniques. First the well-known Principal Component Analysis is introduced, which tries to preserve the variance of the data. Among distance-based mapping methods Sammon mapping and the metric multidimensional scaling will be presened. Multidimensional scaling has a non-metric variant as well, which emphasise the preservation of the rank ordering among the dissimilarities of objects. Sect. 3.3.3 contains the introduction of non-metric variant of multidimensional scaling.

Naturally, there are several other widespread dimensionality reduction methods that provides useful help to decrease the dimensionality of the data points (for example Locally Linearly Embedding [24, 25], Linear Discriminant Analysis [21, 34] or Laplacian Eigenmaps [26]). As these methods do not utilise graphs we do not detail them in this book.

3.3.1 Principal Component Analysis

One of the most widely applied dimensionality reduction methods is the *Principal Component Analysis* (PCA) [16, 17]. The PCA algorithm is also known as Hotteling or as Karhunen-Loéve transform ([16, 17]). PCA differs from the metric and non-metric dimensionality reduction methods, because instead of the preservation of the distances or the global ordering relations of the objects it tries to preserve the variance of the data. PCA represents the data as linear combinations of a small number of basis vectors. This method finds the projection that stores the largest variance possible in the original data and rotates the set of the objects such that the maximum variability becomes visible. Geometrically, PCA transforms the data into a new coordinate system such that the greatest variance by any projection of the data comes to lie on the first coordinate, the second greatest variance on the second coordinate, and so on. If the data set ($\mathbf{X}$) is characterised with D dimensions and the aim of the PCA is to find the d-dimensional reduced representation of the data set, the PCA works as follows: The corresponding d-dimensional output is found by linear transformation: $\mathbf{Y} = \mathbf{QX}$, where $\mathbf{Q}$ is the $d \times D$ matrix of linear transformation composed of the d largest eigenvectors of the covariance matrix, and $\mathbf{Y}$ is the $d \times D$ matrix of the projected data set.

To illustrate PCA-based dimensionality reduction visualisation we have chosen 2 well known data sets. Figure 3.1 shows the PCA-based mapping of the widely

---

Algorithm 9 PCA algorithm

Step 1   PCA subtracts the mean from each of the data dimensions.
Step 2   The algorithm calculates the $D \times D$ covariance matrix of the data set.
Step 3   In the third step PCA calculates the eigenvectors and the eigenvalues of the covariance
matrix.
Step 4   The algorithms chooses the d largest eigenvectors.
Step 5   And finally it derives the new data set from the significant eigenvectors and from the
original data matrix.

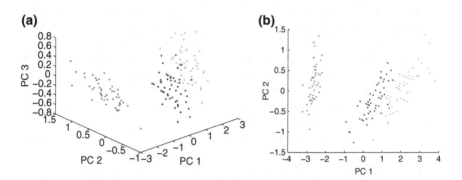

Fig. 3.1 PCA-based visualisations of iris data set. **a** 3-dimensional PCA-based visualisation of iris
data set. **b** 2-dimensional PCA-based visualisation of iris data set

used iris data set (see Appendix A.6.1). In the original data set each sample flower is
characterised with 4 attributes (sepal length and width and petal length and width).
As each sample flower is characterised with 4 numeric values, the original data set
is placed in a 4-dimensional vector space, which is not visible for human eyes. By
the use of PCA this dimensionality can be reduced. In the firs subfigure the first
three principal components are shown on axes ($PC1$, $PC2$, $PC3$), so it provides
a 3-dimensional presentation of the iris data set. In the second subfigure only the
first and the second principal components are shown on axes, therefore in this case
a 2-dimensional visualisation is presented. In these figures each colored plot yields
a flower corresponding to the original data set. Red points indicate iris flowers from
class iris setosa, blue points yield sample flowers from class iris versicolor and
magenta points indicate flowers from class iris virginica.

Figure 3.2 presents the colored S curve data set with 5000 sample points and
its 2-dimensional PCA-based mapping result. The second part of this figure (Fig.
3.2b) demonstrates the main drawback of the Principal Component Analysis based
mapping. As this subfigure do not tarts with dark blue points and do not ends with
bourdon data points, so this method is not able to unfold such linear manifolds that
are nonlinearly embedded in a higher dimensionality space.

As PCA is a linear dimensionality reduction method it can not unfold low-
dimensional manifolds embedded into the high-dimensional vector space. *Kernel
PCA* ([35, 36]) extends the power of the PCA algorithm with applying a kernel trick.

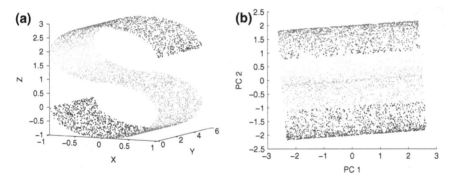

Fig. 3.2 Colored S curve data set and its PCA based mapping. **a** Original S curve data set with $N = 5000$ points. **b** PCA-based visualisation of S curve data set

First it transforms data into a higher-dimensional feature space, and the principal components are in this feature space extracted.

3.3.2 Sammon Mapping

Sammon mapping (SM) [18] is one of the well known metric, nonlinear dimensionality reduction methods. While PCA attempts to preserve the variance of the data during the mapping, Sammon's mapping try to preserve the interpattern distances [37, 38] as it tries to optimise a cost function that describes how well the pairwise distances in a data set are preserved. The aim of the mapping process is to minimise this cost function step by step. The Sammon stress function (distortion of the Sammon projection) can be written as:

$$E_{\text{SM}} = \frac{1}{\sum\limits_{i<j} d^*_{i,j}} \sum_{i<j}^{N} \frac{(d^*_{i,j} - d_{i,j})^2}{d^*_{i,j}}, \tag{3.6}$$

where $d^*_{i,j}$ denotes the distance between the vectors $\mathbf{x}_i$ and $\mathbf{x}_j$, and $d_{i,j}$ respectively for $\mathbf{y}_i$ and $\mathbf{y}_j$.

The minimisation of the Sammon stress is an optimisation problem. When the gradient-descent method is applied to search for the minimum of Sammon stress, a local minimum can be reached. Therefore a significant number of runs with different random initialisations may be necessary.

Figure 3.3 represents 2-dimensional visualisation of the S curve data set resulted by the Sammon mapping. Similar to the previously presented PCA-based mapping, the Sammon mapping can not unfold the nonlinearly embedded 2-dimensional manifold.

Fig. 3.3 Sammon mapping
of S curve data set

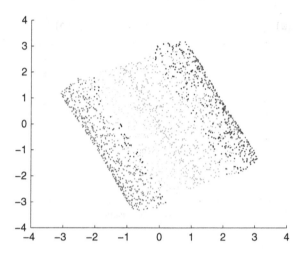

3.3.3 Multidimensional Scaling

Multidimensional scaling (MDS) [27] refers to a group of unsupervised data visualisation techniques. Given a set of data in a high-dimensional feature space, MDS maps them into a low-dimensional (generally 2-dimensional) data space in a way that objects that are very similar to each other in the original space are placed near each other on the map, and objects that are very different from each other are placed far away from each other. There are two types of MDS: (i) *metric MDS* and (ii) *non-metric MDS*.

Metric (or classical) MDS discovers the underlying structure of data set by preserving similarity information (pairwise distances) among the data objects. Similarly to the Sammon mapping the metric multidimensional scaling also tries to minimise a stress function. If the square-error cost is used, the objective function (stress) to be minimized can be written as:

$$E_{\text{metric\_MDS}} = \frac{1}{\sum\limits_{i<j}^{N} d_{i,j}^{*2}} \sum\limits_{i<j}^{N}(d_{i,j}^{*} - d_{i,j})^2, \tag{3.7}$$

where $d_{i,j}^{*}$ denotes the distance between the vectors $\mathbf{x}_i$ and $\mathbf{x}_j$, and $d_{i,j}$ between $\mathbf{y}_i$ and $\mathbf{y}_j$ respectively. The only difference between the stress functions of the Sammon mapping (see 3.6) and the metric MDS (see 3.7) is that the errors in distance preservation in the case of Sammon mapping are normalized by the distances of the input data objects. Because of this normalisation the Sammon mapping emphasises the preservation of small distances.

Classical MDS is an algebraic method that rests on the fact that matrix $\mathbf{Y}$ containing the output coordinates can be derived by eigenvalue decomposition from the scalar product matrix $\mathbf{B} = \mathbf{Y}\mathbf{Y}^T$. Matrix $\mathbf{B}$ can be found from the known distances using Young-Householder process [39]. The detailed metric MDS algorithm is the following:

Metric MDS Algorithm [40]

1. Let the searched coordinates of n points in a d-dimensional Euclidean space be given by $\mathbf{y}_i$ ($i = 1, \ldots, n$), where $\mathbf{y}_i = \left[y_{i,1}, \ldots, y_{i,d}\right]^T$. Matrix $\mathbf{Y} = \left[\mathbf{y}_1, \ldots, \mathbf{y}_n\right]^T$ is the $n \times d$ coordinates matrix. The Euclidean distances $\{d_{i,j} = \left(\mathbf{y}_i - \mathbf{y}_j\right)^T \left(\mathbf{y}_i - \mathbf{y}_j\right)\}$ are known. The inner product of matrix $\mathbf{Y}$ is denoted $\mathbf{B} = \mathbf{Y}\mathbf{Y}^T$. Find matrix $\mathbf{B}$ from the known distances $\{d_{i,j}\}$ using Young-Householder process [39]:

   (a) Define matrix $\mathbf{A} = [a_{i,j}]$, where $a_{i,j} = -\frac{1}{2}d_{i,j}^2$,
   (b) Deduce matrix $\mathbf{B}$ from $\mathbf{B} = \mathbf{H}\mathbf{A}\mathbf{H}$, where $\mathbf{H} = \mathbf{I} - \frac{1}{n}\mathbf{ll}^T$ is the centering matrix, and $\mathbf{l}$ is an $n \times 1$ column vector of n one's.

2. Recover the coordinates matrix $\mathbf{Y}$ from $\mathbf{B}$ using the spectral decomposition of $\mathbf{B}$:

   (a) The inner product matrix $\mathbf{B}$ is expressed as $\mathbf{B} = \mathbf{Y}\mathbf{Y}^T$. The rank of $\mathbf{B}$ is $r(\mathbf{B}) = r\left(\mathbf{Y}\mathbf{Y}^T\right) = r(\mathbf{Y}) = d$. $\mathbf{B}$ is symmetric, positive semi-definite and of rank d, and hence has d non-negative eigenvalues and $n - d$ zero eigenvalues.
   (b) Matrix $\mathbf{B}$ is now written in terms of its spectral decomposition, $\mathbf{B} = \mathbf{V}\Lambda\mathbf{V}^T$, where $\Lambda = \text{diag}[\lambda_1, \lambda_2, \ldots, \lambda_n]$ the diagonal matrix of eigenvalues λ_i of $\mathbf{B}$, and $\mathbf{V} = [\mathbf{v}_1, \ldots, \mathbf{v}_n]$ the matrix of corresponding eigenvectors, normalized such that $\mathbf{v}_i^T \mathbf{v}_i = 1$,
   (c) Because of the $n - d$ zero eigenvalues, $\mathbf{B}$ can now be rewritten as $\mathbf{B} = \mathbf{V}_1\Lambda_1\mathbf{V}_1^T$, where $\Lambda_1 = \text{diag}[\lambda_1, \lambda_2, \ldots, \lambda_d]$ and $\mathbf{V}_1 = [\mathbf{v}_1, \ldots, \mathbf{v}_d]$,
   (d) Finally the coordinates matrix is given by $\mathbf{Y} = \mathbf{V}_1\Lambda_1^{\frac{1}{2}}$, where $\Lambda_1^{\frac{1}{2}} = \text{diag}\left[\lambda_1^{\frac{1}{2}}, \ldots, \lambda_d^{\frac{1}{2}}\right]$.

To illustrate the method visualisation of the iris data set and points lying on an S curve were chosen (see Fig. 3.4).

In contrast with metric multidimensional scaling, in *non-metric MDS* only the ordinal information of the proximities is used for constructing the spatial configuration. Thereby non-metric MDS attempts to preserve the rank order among the dissimilarities. The non-metric MDS finds a configuration of points whose pairwise Euclidean distances have approximately the same rank order as the corresponding dissimilarities of the objects. Equivalently, the non-metric MDS finds a configuration of points, whose pairwise Euclidean distances approximate a monotonic transforma-

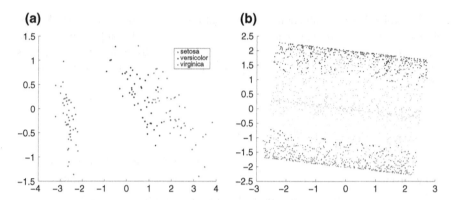

Fig. 3.4 MDS mappings of iris and S curve data sets. **a** MDS mapping of iris data set. **b** MDS mapping of S curve data set

tion of the dissimilarities. These transformed values are known as the disparities. The non-metric MDS stress can be formulated as follows:

$$E_{\text{nonmetric\_MDS}} = \sqrt{\sum_{i<j}^{N} (\widehat{d}_{i,j} - d_{i,j})^2 / \sum_{i<j}^{N} d_{i,j}^2}, \qquad (3.8)$$

where $\widehat{d}_{i,j}$ yields the disparity of objects $\mathbf{x}_i$ and $\mathbf{x}_j$, and $d_{i,j}$ denotes the distance between the vectors $\mathbf{y}_i$ and $\mathbf{y}_j$. Traditionally, the non-metric MDS stress is often called Stress-1 due to Kruskal [41].

The main steps of the non-metric MDS algorithm are given in Algorithm 10.

Algorithm 10 Non-metric MDS algorithm

Step 1 Find a random configuration of points in the output space.
Step 2 Calculate the distances between the points.
Step 3 Find the optimal monotonic transformation of the proximities in order to obtain the disparities.
Step 4 Minimise the non-metric MDS stress function by finding a new configuration of points.
Step 5 Compare the stress to some criteria. If the stress is not enough small then go back to Step 2.

It can be shown, that metric and non-metric MDS mappings are substantially different methods. On the one hand, while metric MDS algorithm is an algebraic method, the non-metric MDS is an iterative mapping process. On the other hand the main goal of the optimisation differs significantly, too. While metric multidimensional scaling methods attempt to maintain the degree of the the pairwise dissimilarities of data points, the non-metric multidimensional scaling methods focus on the preservation of the order of the neighbourhood relations of the objects.

3.4 Neighbourhood-Based Dimensionality Reduction

In this session dimensionality reduction methods utilising topology of data in the input space are presented.

3.4.1 Locality Preserving Projections

Locality Preserving Projections (LPP) [42] method is a linear dimensionality reduction method, which can be seen as a linear extension of Laplacian eigenmaps. First, the algorithm builds a graph based on k-neighbouring or ε-neighbouring in the input space and during the mapping it tries to preserve optimally the neighbourhood structure of the data. The mapping component of the algorithm is a linear approximation to the eigenfunctions of the Laplace Beltrami operator on the manifold embedded in the high-dimensional vector space. The detailed algorithm can be found in Algorithm 11.

Algorithm 11 Locality Preserving Projections algorithm

Given a set of input objects $\mathbf{X} = \{\mathbf{x}_1, \mathbf{x}_2, \ldots, \mathbf{x}_N\}^T, \mathbf{x}_i \in \mathbb{R}^D, i = 1, 2, \ldots, N$.

Step 1   In the first step the adjacency graph of the objects is calculated. The algorithm connects two objects with an edge if they are close to each other. The closeness is determined based on the of k-neighbouring or ε-neighbouring approaches.

Step 2   In the second step weights are assigned to the edges. The weights can be calculated based on the following two principle:

Heat kernel: For each edges connecting $\mathbf{x}_i$ and $\mathbf{x}_j$ the weight is calculated as follows:

$$w_{ij} = e^{-\frac{\|\mathbf{x}_i - \mathbf{x}_j\|^2}{t}}, \tag{3.9}$$

where t is an input parameter.

Simple-minded: $w_{ij} = 1$ if objects $\mathbf{x}_i$ and $\mathbf{x}_j$ are connected by an edge, otherwise $w_{ij} = 0$.

Step 3   In the third step the algorithm computes the eigenvectors and eigenvalues for the following generalized eigenvector problem:

$$\mathbf{XLX}^T\mathbf{a} = \lambda \mathbf{XMX}^T\mathbf{a}, \tag{3.10}$$

where $\mathbf{M} = \mathbf{L} - \mathbf{W}$ is the Laplacian matrix, and $\mathbf{M}$ is a diagonal matrix where $m_{ii} = \sum_j w_{ji}$. $\mathbf{X}$ is the data matrix and the ith column of that matrix is the ith object $\mathbf{x}_i$.

Column vectors $\mathbf{a}_0, \mathbf{a}_1, \ldots \mathbf{a}_N - 1$ are the solutions of Equation 3.10, ordered according to the eigenvectors $\lambda_0 < \lambda_1 < \ldots < \lambda_N - 1$. The mapped data points $\mathbf{y}_i$ ($i = 1, 2, \ldots, N$) are calculated as follows:

$$\mathbf{y}_i = \mathbf{A}^T\mathbf{x}_i, \tag{3.11}$$

where $\mathbf{A} = (\mathbf{a}_0, \mathbf{a}_0, \ldots, \mathbf{a}_{N-1})$.

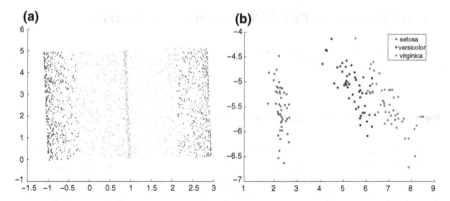

Fig. 3.5 LPP mappings of S curve and iris data sets. **a** LPP mapping of S curve data set. **b** LPP mapping of iris data set

As a result LPP provides linear dimensionality reduction, which aims to preserve the local structure of a given data set. Running experiments show, that LPP is not sensitive to outliers and noise [42].

Figures 3.5 and 3.6 illustrate three application examples. Beside mapping of Iris and S curve data sets (Fig. 3.5) a new benchmark data set is also selected for the visualisation. The Semeion data set (see Appendix A.6.2 [43]) contains 1593 handwritten

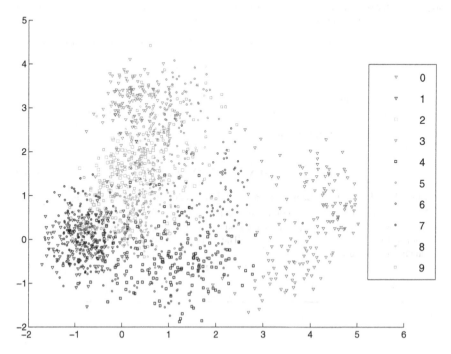

Fig. 3.6 LPP mapping of semeion data set

digits from around 80 persons. Each person wrote on a paper all the digits from 0 to 9, twice. First time in the normal way as accurate as they can and the second time in a fast way. The digits were scanned and stretched in a rectangular box including 16×16 cells in a grey scale of 256 values. Then each pixel of each image was scaled into a boolean value using a fixed threshold. As a result the data set contains 1593 sample digits and each digit is characterised with 256 boolean variables. LPP of the Semeion data set is shown in Fig. 3.6. The resulted 2-dimensional visualisation shows interesting correlations between the digits.

3.4.2 Self-Organizing Map

Nowadays, neural networks (NN) [44] are also widely used to reduce dimensionality. *Self-Organizing Map* (SOM) developed by professor Teuvo Kohonen in the early 1980s [23] is one of the most popular neural network models. The main goal of SOM is to transform a high-dimensional (D-dimensional) pattern into a low-dimensional discrete map in a topologically ordered grid. The 2-dimensional grids may be arranged in a rectangular or a hexagonal structure. These fixed grid structures are shown in Fig. 3.7. Although SOM can handle missing values, it is not able to preserve the topology of the input data structure, when map structure does not match the input data structure. Actually, the SOM is a neural network, where each neuron is represented by a D-dimensional weight vector.

SOM is trained iteratively, using unsupervised learning. During the training process the weight vectors have to be initialised first (e.g. randomly or sample initialisation). After that given a random sample from the training data set, the best matching unit (BMU) in the SOM is located. The BMU is the closest neuron to the selected input pattern based on the Euclidean distance. The coordinates of BMU and neurons closest to it in the SOM grid are then updated towards the sample vector in the input space. The coverage of the change decreases with time. BMU and its neighbouring neurons in the SOM grid are updated towards the sample object based on the following formula:

$$\mathbf{w}_i (t + 1) = \mathbf{w}_i (t) + h_{\text{BMU},i} (t) [\mathbf{x} (t) - \mathbf{w}_i (t)], \qquad (3.12)$$

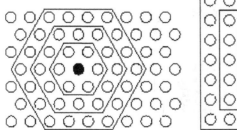 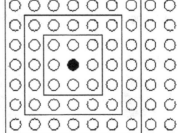

Fig. 3.7 Hexagonal and rectangular structure of SOM

where t denotes time, $\mathbf{w}_i$ denotes the neurons in the grid, $\mathbf{x}(t)$ is the random sample object at time t and $h_{c,i}(t)$ yields the neighbourhood function about the winner unit (BMU) at time t.

The training quality of the Self-Organizing Map may be evaluated by the following formula:

$$E_{\text{SOM}} = \frac{1}{N} \sum_{i=1}^{N} \|\mathbf{x}_i - \mathbf{w}_{\text{BMU}}^i\|, \tag{3.13}$$

where N is the number of the objects to be mapped and $\mathbf{w}_{\text{BMU}}^i$ yields the best matching unit corresponding to the vector $\mathbf{x}_i$.

When SOM has been trained, it is ready to map any new input vector into a low-dimensional vector space. During the mapping process a new input vector may quickly be classified or categorized, based on the location of the closest neuron on the grid.

There is a variety of different kinds of visualisation techniques available for the SOM. (e.g. U-matrix, component planes). The *Unified distance matrix* (U-matrix) [45] makes the 2D visualisation of multi-variate data. In the U-matrix the average distances between the neighbouring neurons are represented by shades in a grey scale. If the average distance of neighbouring neurons is short, a light shade is used; dark shades represent long distances. Thereby, dark shades indicate a cluster border, and light shades represent clusters themselves. *Component planes* [23] are visualised by taking from each weight vector the value of the component (attribute) and depicting this as a color or as a height on the grid. Figure 3.8 illustrates the U-matrix and the

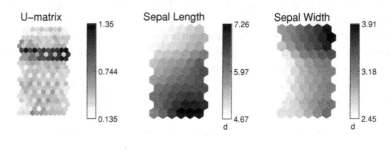

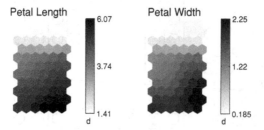

Fig. 3.8 The U-matrix and the component planes of the iris data set

component planes representation in a hexagonal grid of the well-known iris data set. It can be seen that the first iris kind (Iris setosa) forms a separate cluster and the second kind (Iris versicolor) and the third one (Iris virginica) are mixed a little bit. Among the wide range of other possibilities, the Self-Organizing Map is often visualised by Principal Component Analysis and Sammon mapping to give more insight to the structure of high-dimensional data.

3.4.3 Incremental Grid Growing

Incremental Grid Growing (IGG) [46] is a neural network based dimensionality reduction approach, which takes into account the distribution of analysed data. The algorithm results in a 2-dimensional grid that explicitly represents the cluster structure of the high-dimensional input data elements. Namely, different clusters are represented by separated grids.

Opposite to the SOM, the IGG algorithm is an incremental method, which step by step adds nodes to the resulted 2-dimensional graph in the output space. As a consequence, the IGG algorithm does not require to define the size of the map a priori. Furthermore, the edges of the nodes are created and deleted dynamically, too. Analogously to the SOM, the Incremental Growing Grid algorithm results in a fixed grid topology, as well. While the SOM uses a predefined rectagonal or hexagonal data structure in the the reduced space, the IGG algorithm works with a regular 2-dimensional grid. This 2-dimensional grid allows to define the boundary nodes of the grid, which is a decisive concept of the Incremental Grid Growing algorithm. In Blackmore's and Miikkulainen' article the boundary node is defined as a node in the grid that has at last one directly neighbouring position in the grid not yet occupied by a node [46].

The Incremental Growing Grid algorithm starts with 2×2 nodes in the output space. All boundary nodes are characterised by an error variable initialised to zero. The algorithm works iteratively, in each iteration it chooses an input data object randomly. Each iteration consist of three phases. During the *training phase* nodes in the output grid are adapted to the input data elements in the same way as it in the SOM algorithm happens. In the *expansion phase* the IGG algorithm creates a new node or new nodes if it is necessary. If the closest node in the output grid to the arbitrary chosen input vector is an boundary node, the algorithm recalculates the error variable of this boundary node as follows:

$$E_i(t) = E_i(t-1) + \sum_k \left(\mathbf{x}_k - \mathbf{w}_{i_k}\right)^2, \tag{3.14}$$

where t refers to the current number of iterations, $\mathbf{w}_i$ is the winner boundary node, and $\mathbf{x}$ is the arbitrary selected input object. The algorithm adds new nodes to the net if too large number of analysed data are mapped into a low-dimensional output node, and therefore the cumulative error of this node is too large. The boundary node with

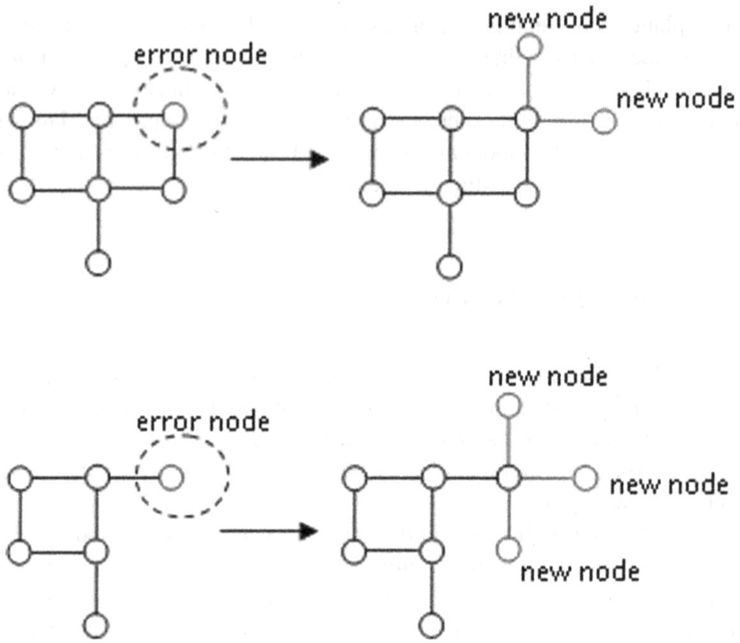

Fig. 3.9 Adding new nodes to the grid

the largest cumulative error (error node) is selected as the most inadequate node to represent the data structure. This node grows new neighbouring node or nodes. The new nodes are placed in all possible position in the grid which are adjacent to the winner boundary node and they are not yet occupied by a node. Figure 3.9 visualises two kinds of growing possibilities.

Weight vectors of the new nodes are initialised based on the neighbouring nodes and the new nodes are connected to the error node. Finally, during the *adaptation of connections phase* the algorithm evaluates two parameters for each connection. If the Euclidean distance between two neighbouring unconnected nodes are less than a connect threshold parameter, the algorithm creates a new edge between these nodes. On the other hand, if the distance between two nodes connected in the grid exceeds a disconnect threshold parameter, the algorithm deletes it.

To summarise, we can see that the Incremental Grid Growing algorithm analogously to the SOM method utilises a predefined 2-dimensional structure of representative elements, but in the case of the IGG algorithm the number of these nodes is not a predefined parameter. As a consequence of the deletion of edges the IGG algorithm may provide unconnected subgrids, which can be seen as a representation of different clusters of the original objects.

The *Adaptive Hierarchical Incremental Grid Growing* (AHIGG) method proposed by Merkl and coworkers in 2003 [47] extends the Incremental Grid Growing approach. In this article the authors combine the IGG method with the hierarchical clustering approach. The main difference between the IGG and AHIG method is, that in the course of the AHIGG algorithm the network representing the data grows incrementally, but there are different levels of the growing state distinguished. The Adaptive Hierarchical Incremental Grid Growing algorithm utilises the SOM algorithm to train the net as well, but the initialisation of the net differs from the method proposed in the IGG algorithm. The training process involves a fine tuning phase as well, when only the winner node adapts to the selected data point, and no further nodes are added to the graph. After the fine tuning phase the algorithm searches for the possible extensions in the graph. For this extension the algorithm calculates an error value (mean quantisation error) for each node, and nodes with too high error value are expanded on the next level of the presentation. As a result the algorithm creates a hierarchical architecture of different visualisation levels. Each level of the hierarchy involves a number of independent clusters presented by 2-dimensional grid structures.

3.5 Topology Representation

The main advantage of low-dimensional visualisation is that in the low-dimensional vector space human eyes are able to detect clusters, or relationships between different data objects or group of data. This is basic idea of the exploratory data analysis. Methods introduced so far do not give back graphically the real structure of data. Either they are not able to unfold manifolds embedded nonlinearly into the high-dimensional space, or they visualise the observed objects or its representatives in a predefined structure like a grid, or hexagonal lattice.

In this subsection such algorithms are introduced which try to visualise the real, sometimes hidden structure of data to be analysed. These methods are mainly based on the concept of geodesic distance. The geodesic distance is more appropriate choice to determine the distances of the original objects or its representatives that the Euclidean distance, as it calculates the pairwise distances along the manifold. Therefore these algorithm are able to unfold such lower dimensional manifolds as well which are nonlinearly embedded into the high-dimensional vectorspace.

The calculation of the geodesic distances of the objects requires the creation of a graph structure of the objects or its representatives. This network may be established for example by connecting the k nearest neighbours, or based on the principle of ε-neighbouring. The well-known Isomap [19], Isotop [23] and Curvilinear Distance Analysis [48] methods utilise all the k-neighbouring or ε-neighbouring approaches to describe the neighbourhood-based structure of the analysed data. These methods are introduced in Sects. 3.5.1–3.5.3.

Other possibility is to create topology representing networks to capture the data structure. In the literature only few methods are only published that utilise the topology representing networks to visualise the data set in the low-dimensional vector space. The Online visualisation Neural Gas (OVI-NG) [49] is a nonlinear projection method, in which the mapped representatives are adjusted in a continuous output space by using an adaptation rule that minimises a cost function that favors the local distance preservation. As OVI-NG utilises Euclidean distances to map the data set it is not able to disclose the nonlinearly embedded data structures. The Geodesic Non-linear Projection Neural Gas (GNLP-NG) [50] algorithm is an extension of OVI-NG, which uses geodesic distances instead of the Euclidean ones. The TRNMap algorithm was developed recently, and it combines the TRN-based geodesic distances with the multidimensional scaling method. In Sects. 3.5.4–3.5.6 these algorithms are introduced.

3.5.1 Isomap

The *Isomap* algorithm proposed by Tenenbaum et al. in 2000 [19] is based on the geodesic distance measure. Isomap deals with finite number of points in a data set in $\mathbb{R}^D$ which are assumed to lie on a smooth submanifold M^d ($d \ll D$). The aim of this method is to preserve the intrinsic geometry of the data set and visualise the data in a low-dimensional feature space. For this purpose Isomap calculates the geodesic distances between all data points and then projects them into a low-dimensional vector space. In this way the Isomap algorithm consists of three major steps:

Step 1 : Constructing the neighbourhood graph of the data by using the k-neighbouring or ε-neighbouring approach.
Step 2 : Computing the geodesic distances between every pair of objects.
Step 3 : Constructing a d-dimensional embedding of the data points.

For the low-dimensional (generally $d = 2$) visualisation Isomap utilises the MDS method. In this case the multidimensional scaling is not based on the Euclidean distances, but it utilises the previously computed geodesic distances. As Isomap uses a non-Euclidean metric for mapping, a nonlinear projection is obtained as a result.

However, when the first step of the Isomap algorithm is applied to a multi-class data set, several disconnected subgraphs can be formed, thus the MDS can not be performed on the whole data set. Wu and Chan [51] give an extension of the Isomap solving this problem. In their proposal unconnected subgraphs are connected with an edge between the two nearest node. In this manner the Euclidean distance is used to approximate the geodesic distances of data objects lying on different disconnected subgraphs. Furthermore, applying Isomap to noisy data shows also some limitations.

Figures 3.10 and 3.11 illustrate two possible 2-dimensional Isomap mappings of the S curve set. It can be seen that due to the calculation of the geodesic distances, the Isomap algorithm is able to unfold the 2-dimensional manifold nonlinearly embedded into the 3-dimensional vector space. Additionally, in this figures two different mapping results can be seen, which demonstrate the effect of the parametrisation of Isomap algorithm. To calculate geodesic distances the k-neighbouring approach was chosen in both cases. In the first case k was chosen to be $k = 5$, in the second case k was set to be $k = 10$.

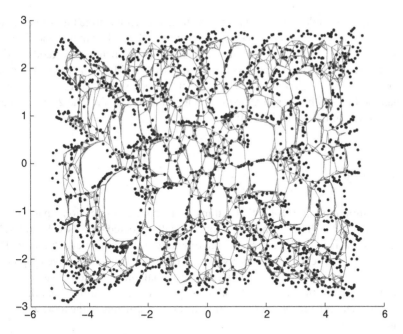

Fig. 3.10 Isomap mapping of S curve data set ($k = 5$)

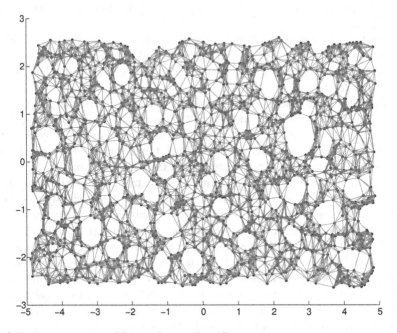

Fig. 3.11 Isomap mapping of S curve data set ($k = 10$)

As second example the iris data set was chosen to demonstrate the Isomap mapping. In this case the creation of the neighbouring graph resulted in two unconnected subgraphs and so the original algorithm was not able to calculate the geodesic distances between all pairs of data. As a consequence, the MDS mapping can not be performed on the whole data set. Therefore it can be established that the original Isomap method (without any extensions) is not able to visualise the 2-dimensional representation of the whole iris data set.

3.5.2 Isotop

The main limitation of SOM is that it transforms the high-dimensional pattern into a low-dimensional discrete map in a topologically ordered grid (see Sect. 3.4.2). Thereby, SOM is not able to preserve the topology of the input data structure. The method *Isotop* [52] can be seen as a variation of SOM with a data-driven topology grid. Contrary to the SOM's rectangular or hexagonal lattice, Isotop creates a graph that tries to capture the neighbourhood relationships in the manifold, and therefore the resulted network reflects more accurate the hidden structure of the representatives or data elements.

The algorithm consist of 3 major steps: (1) vector quantisation; (2) building a graph from the representative elements; (3) mapping the graph onto a low-dimensional vector space.

In the first phase Isotop performs a vector quantisation step in order to reduce the number of data points. So the objects are replaced by their representatives. This optional step can be achieved with simple methods, like Competitive Learning or k-means clustering.

In the second step Isotop builds a graph structure to calculate the geodesic distances of the objects. This network is created based on the k-neighbouring or ε-neighbouring approaches. Parameters k or ε are determined by the analyser. In the network the edges are characterised by the Euclidean distances of the objects, and the geodesic distances are calculated as sums of these Euclidean distances.

Finally, in the last step Isotop performs a non-metric dimensionality reduction. This mapping process uses graph distances defined by the previously calculated neighbourhood connections. Up to this point the analysed objects are represented with representative elements in the high-dimensional vector space. In this step the algorithm replaces the coordinates of representatives by low-dimensional ones, initialised randomly around zero. Then Isotop iteratively draws an object ($\mathbf{g}$) randomly, and moves all representatives closer to the randomly selected point. The movement of each mapped representative becomes smaller and smaller as its neighbourhood distance from the closest representative to the selected point (BMU) grows. Formally, at time t all representatives $\mathbf{y}_i$ in the low-dimensional space are updated according to the rule:

$$\mathbf{y}_i = \mathbf{y}_i + \alpha(t)h_i(t)(\mathbf{g}(t) - \mathbf{y}_i), \tag{3.15}$$

where $\alpha(t)$ is a time-decreasing learning rate with values taken from between 1 and 0. The neighbourhood factor h_i is defined as:

$$h_i(t) = \exp\left(-\frac{1}{2} \frac{\delta_{i,j}^2}{\left(\lambda(t) E_{j \in N(i)} \{\|\mathbf{x}_i \mathbf{x}_j\|\}\right)^2}\right), \tag{3.16}$$

where $\lambda(t)$ is a time-decreasing neighbourhood width, $\delta_{i,j}$ is the graph distance between the objects $\mathbf{x}_i$ and $\mathbf{x}_j$, and $E_{j \in N(i)} \{\|\mathbf{x}_i \mathbf{x}_j\|\}$ is the mean Euclidean distance between the i-th representative element and its topological neighbours.

To summarise, we can say that Isotop tries to preserve the neighbourhood relations of the representatives in the low-dimensional output space. The algorithm is able to unfold low-dimensional manifolds nonlinearly embedded into a high-dimensional vector space, but it is sensitive to the parametrisation and it may fall in local minima [53].

3.5.3 Curvilinear Distance Analysis

Curvilinear Component Analysis (CCA) [54] was also proposed as an improvement of the Kohonen's Self-Organizing Maps, in that the output lattice has no fixed structure predetermined. *Curvilinear Distance Analysis* (CDA) [48] is the nonlinear variant of CCA. While CCA is based on the Euclidean distances, CDA utilises the curvilinear (graph) distances.

CDA algorithm performs the following five tasks separately:

Step 1 : Vector quantisation (optional step).
Step 2 : Computing the k- or ε-neighbourhoods and linking of the representatives (or objects).
Step 3 : Calculating the geodesic distances by Dijkstra's algorithm (see Appendix A.2.1).
Step 4 : Optimizing a cost function by stochastic gradient descent, in order to get coordinates for the representatives (or objects) in the output space.
Step 5 : After mapping of the representatives the full data set is mapped by running a piecewise linear interpolator (this step is unnecessary if step 1 was skipped).

In the 4-th step CDA maps the objects or its representatives by minimizing the topology error function defined as:

$$E_{\text{CDA}} = \sum_{i<j}^{n} \left(\delta_{i,j} - d_{i,j}\right)^2 F\left(d_{i,j}, \lambda\right), \tag{3.17}$$

where $\delta_{i,j}$ denotes the geodesic distance between the objects $\mathbf{x}_i$ and $\mathbf{x}_j$ in the high-dimensional input space, $d_{i,j}$ denotes the Euclidean distance for the mapped objects $\mathbf{y}_i$ and $\mathbf{y}_j$ in the low-dimensional output space, and n is the number of the

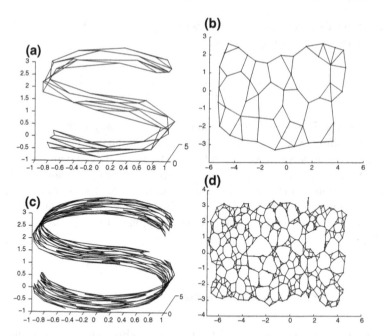

Fig. 3.12 Nearest neighbours graphs and CDA mappings of quantised S curve data set. **a** Nearest neighbours graph of S curve data set $n = 50$, $k = 3$. **b** CDA mapping of S curve data set $n = 50$. **c** Nearest neighbours graph of S curve data set $n = 500$, $k = 3$. **d** CDA mapping of S curve data set $n = 500$

representatives (or objects) to be mapped. F is a decreasing function of $d_{i,j}$ ($F : \mathbb{R}^+ \to [0, 1]$), and λ is the neighbourhood radius. The function F allows the local topology to be favored with respect to the global topology. Usually, F is implemented as the Heaviside step function:

$$F\left(d_{i,j}, \lambda\right) = \begin{cases} 0, & \text{if } \lambda - d_{i,j} < 0, \\ 1, & \text{if } \lambda - d_{i,j} \geq 0. \end{cases} \tag{3.18}$$

The application of factor F effects that CDA algorithm emphasises the preservation of small distances rather than of large ones. Curvilinear Distance Analysis applies stochastic gradient descent algorithm to minimise the topology error function E_{CDA}.

Figure 3.12 demonstrates some CDA mappings of S curve data set. In both cases vector the process of the quantisation was made based on the k-means algorithm where the number of the representatives in the first case was chosen to be $n = 100$ and in the second case $n = 500$. The number of the original data points in all cases were $N = 2000$. In the left column the k nearest neighbours graphs are shown where the number of the nearest neighbours in all cases was chosen to be $k = 3$. The right column contains the CDA mappings of the vector quantised data.

Comparing Isomap and CDA methods it can be seen, that CDA applies more complicated techniques than the Isomap. However, when the parametrisation is adequate, CDA may give better visualisation result, which emphasise better some characteristics of the projected data sets [55].

3.5.4 Online Data Visualisation Using Neural Gas Network

Online visualisation Neural Gas (OVI-NG) algorithm [49] is a nonlinear projection method which combines the TRN algorithm with an adaptation rule to establish the codebook positions ($\mathbf{Y} = \{\mathbf{y}_1, \mathbf{y}_2, \ldots, \mathbf{y}_n\}$, $\mathbf{y}_i \in \mathbb{R}^d$, $i = 1, \ldots, n$) in the low-dimensional output space. Codebook positions mean the mapped codebook vectors in the low-dimensional output space. The method adjusts codebook vectors in the input space and their respective codebook positions in the output space simultaneously. To obtain a distance preserving mapping, the OVI-NG defines the cost function as follows:

$$E_{\text{OVI-NG}} = \frac{1}{2} \sum_{j=1}^{n} \sum_{k \neq j} (d_{j,k}^* - d_{j,k})^2 F(s_{j,k}), \qquad (3.19)$$

where $d_{j,k}^*$ defines the Euclidean distance in the input space between the codebook vectors $\mathbf{w}_j$ and $\mathbf{w}_k$, $d_{j,k}$ yields the Euclidean distance of the codebook positions $\mathbf{y}_j$ and $\mathbf{y}_k$ in the output space, and $s_{j,k}$ denotes the rank of the k-th codebook position ($\mathbf{y}_k$) with respect to the j-th output vector ($\mathbf{y}_j$) in the output space. The function F is defined as:

$$F(f) = e^{-\left(\frac{f}{\sigma(t)}\right)}, \qquad (3.20)$$

where $\sigma(t)$ is the width of the neighbourhood that decreases with the number of iterations in the same way as Eq. 1.12.

The OVI-NG method performs 10 steps separately. As some of them are equivalent with steps of TRN algorithm, in the following only the additional steps are discussed in detail. **Steps 1–7** in the OVI-NG method are the same as Steps 1–7 in the TRN algorithm (see Sect. 1.2.4), except that in the first step beside the random initialisation of the codebook vectors $\mathbf{w}_j$ the OVI-NG also initialises codebook positions $\mathbf{y}_j$ randomly. In each iteration step after creating new edges and removing the 'old' edges (Step 5–7), the OVI-NG moves the codebook positions closer to the codebook position associated with the winner codebook vector ($\mathbf{w}_{j_0}$). This adaptation rule is carried out by the following two steps:

Step 8 Generate the ranking in output space $s(j_0, j) = s(\mathbf{y}_{j_0}(t), \mathbf{y}_j(t)) \in \{1, \ldots, n-1\}$ for each codebook position $\mathbf{y}_j(t)$ with respect to the codebook position associated with the winner unit $\mathbf{y}_{j_0}(t)$, $j \neq j_0$.

Step 9 Update the codebook positions:

$$\mathbf{y}_j(t+1) = \mathbf{y}_j(t) + \alpha(t) F(s_{j_0,j}) \frac{\left(d_{j_0,j} - d^*_{j_0,j}\right)}{d_{j_0,j}} \left(\mathbf{y}_{j_0}(t) - \mathbf{y}_j(t)\right) \quad (3.21)$$

where α is the learning rate, which typically decreases with the number of iterations t, in the same form as Eq. 1.12.

Step 10 of the OVI-NG is the same as Step 8 in the TRN algorithm.

To sum up, we can say that OVI-NG is a nonlinear projection method, in which the codebook positions are adjusted in a continuous output space by using an adaptation rule that minimises a cost function that favors the local distance preservation. As OVI-NG utilises Euclidean distances to map the data set it is not able to disclose the nonlinearly embedded data structures.

3.5.5 Geodesic Nonlinear Projection Neural Gas

The main disadvantage of the OVI-NG algorithm is that it is based on Euclidean distances, hence it is not able to uncover nonlinearly embedded manifolds. The *Geodesic Nonlinear Projection Neural Gas* (GNLP-NG) [50] is an extension of OVI-NG, which uses geodesic distances instead of the Euclidean ones. The GNLP-NG method includes the following two major steps:

1. creating a topology representing network to depict the structure of the data set and then,
2. mapping this approximate structure into a low-dimensional vector space.

The first step utilises neural gas vector quantisation to define the codebook vectors in the input space, and it uses the competitive Hebbian rule for building a connectivity graph linking these codebook vectors. The applied combination of the neural gas method and the Hebbian rule differs slightly from the TRN algorithm: it connects not only the first and the second closest codebook vectors to the randomly selected input pattern $(\mathbf{x}_i(t))$ (Step 5 in the TRN algorithm (see Algorithm 4 in Sect. 1.2.4)), but it creates other edges, as well. The establishment of these complementary edges can be formalised in the following way:

- for $k = 2, \ldots, K$
  Create a connection between the k-th nearest unit $(\mathbf{w}_{j_k})$ and the $k + 1$-th nearest unit $(\mathbf{w}_{j_{k+1}})$, if it does not exist already, and the following criterion is satisfied:

$$\|\mathbf{w}_{j_k} - \mathbf{w}_{j_{k+1}}\| < \|\mathbf{w}_{j_0} - \mathbf{w}_{j_{k+1}}\| \quad (3.22)$$

Else create a connection between codebook vectors $\mathbf{w}_{j_0}$ and $\mathbf{w}_{j_{k+1}}$.
Set the ages of the established connections to zero. If the connection already exists, refresh the age of the connection by setting its age to zero.

Parameter K is an accessory parameter compared to the TRN algorithm. In [50] it is suggested to set $K = 2$. This accessory step amends the 5-th step of the TRN algorithm (see Sect. 1.2.4).

Furthermore GNLP-NG increments not only the ages of all connections of $\mathbf{w}_{i_0}$, but it also extends this step to the k-th nearest unit as follows:

- Increment the age of all edges emanating from $\mathbf{w}_{i_k}$, for $k = 1, \ldots, K$:

$$t_{i_0,l} = t_{i_0,l} + 1, \forall l \in N_{\mathbf{w}_{i_k}}, \tag{3.23}$$

where $N_{\mathbf{w}_{i_k}}$ is the set of all direct topological neighbours of $\mathbf{w}_{i_k}$.

This accessory step amends the 6-th step of the TRN algorithm (see Algorithm 4).

During the mapping process (second major part of the algorithm) the GNLP-NG algorithm applies an adaptation rule for the codebook positions in the projection space. It minimises the following cost function:

$$E_{\text{GNLP-NG}} = \frac{1}{2} \sum_{j=1}^{n} \sum_{k \neq j} (d_{j,k} - \delta_{j,k})^2 e^{-\left(\frac{\bar{r}_{j,k}}{\sigma(t)}\right)^2}, \tag{3.24}$$

where $\bar{r}_{j,k} = \bar{r}(\mathbf{x}_j, \mathbf{w}_k) \in \{0, 1, \ldots, n-1\}$ denotes the rank of the k-th codebook vector with respect to the $\mathbf{x}_j$ using geodesic distances, and σ is a width of the neighbourhood surround. $d_{j,k}$ denotes the Euclidean distance of the codebook positions $\mathbf{y}_j$ and $\mathbf{y}_k$ defined in the output space, $\delta_{j,k}$ yields the geodesic distance between codebook vectors $\mathbf{w}_j$ and $\mathbf{w}_k$ measured in the input space.

According to the previously presented overview, the GNLP-NG first determines the topology of the data set by the modified TRN algorithm and then maps this topology based on the graph distances. The whole process is summarised in Algorithm 12.

Parameter α is the learning rate, σ is the width of the neighbourhood, and they typically decrease with the number of iterations t, in the same way as Eq. 1.12. Paper [50] also gives an extension to the GNLP-NG to tear or cut the graphs with non-contractible cycles.

Figure 3.13 visualises the 2-dimensional GNLP-NG mapping of the S curve data set. In this small example the original S curve data set contained 2000 3-dimensional data points, the number of the representatives was chosen to be $n = 200$ and they were initialised randomly. As it can be seen the GNLP-NG method is able to unfold the real 2-dimensional structure of the S curve data set.

Algorithm 12 GNLP-NG algorithm

Step 1   Determine the topology of the data set based on the modified TRN algorithm.
Step 2   Compute the geodesic distances between the codebook vectors based on the connections
$(c_{i,j})$ of the previously calculated topology representing network. Set $t = 0$.
Step 3   Initialize the codebook positions $\mathbf{y}_j$, randomly.
Step 4   Select an input pattern $\mathbf{x}_i(t)$ with equal probability for each $\mathbf{x} \in \mathbf{X}$.
Step 5   Find the codebook vector $\mathbf{w}_{j_0}(t)$ in the input space that is closest to $\mathbf{x}_i(t)$.
Step 6   Generate the ranking using geodesic distances in the input space
$\bar{r}_{j_0,j} = \bar{r}(\mathbf{w}_{j_0}(t), \mathbf{w}_j(t)) \in \{0, 1, \ldots, n-1\}$ for each codebook vector $\mathbf{w}_j(t)$ with respect to
$\mathbf{w}_{j_0}(t)$.
Step 7   Update the codebook positions in the output space:

$$\mathbf{y}_j(t+1) = \mathbf{y}_j(t) + \alpha(t)e^{-\left(\frac{\bar{r}_{j_0,j}}{\sigma(t)}\right)^2}\frac{(d_{j_0,j} - \delta_{j_0,j})}{d_{j_0,j}}(\mathbf{y}_{j_0}(t) - \mathbf{y}_j(t)) \qquad (3.25)$$

Step 8   Increase the iteration counter $t = t + 1$. If $t < t_{max}$ go back to Step 4.

Fig. 3.13 GNLP-NG mapping of S curve data set

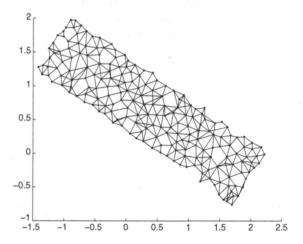

3.5.6 Topology Representing Network Map

Summarising the previously introduced methods we can say, that all these methods seem to be a good choice for topology based dimensionality reduction, but each of them has some disadvantages. Isomap can not model multi-class problems and it is not efficient on large and noisy data sets. The main disadvantage of OVI-NG and GNLP-NG methods are that they use a non-metric mapping method and thereby only the rank ordering of the representatives is preserved during the mapping process. Isotop can indeed fall in local minima and require some care for the parametrisation [53]. Although CDA is a more complicated technique, it needs to be well parameterized [56]. Furthermore, the OVI-NG and CCA methods are not able to uncover the non-linearly embedded manifolds.

Topology Representing Network Map (TRNMap) [57, 58] refers to a group of unsupervised nonlinear mapping methods, which combines the TRN algorithm and the multidimensional scaling to visualise the data structure. As result it gives a compact representation of the data set to be analysed. The method aims to fulfill the following three criteria:

- give a low-dimensional representation of the data,
- preserve the intrinsic data structure (topology), and
- according to the users expectations: preserve the distances or the rank ordering of the objects.

TRNMap mapping method results in a visualisation map, called Topology Representing Network Map (TRNMap). TRNMap is a self-organizing model with no predefined structure which provides an expressive presentation of high-dimensional data in low-dimensional vector space. The dimensionality of the input space is not restricted. Although this method is able to provide arbitrary dimensional output map as result, for the visualisation of data structure the 2-dimensional or 3-dimensional output map is recommended. Topology Representing Network Map algorithm is based on graph distances, therefore it is able to handle the set of data lying on a low-dimensional manifold that is nonlinearly embedded in a higher-dimensional input space. For the preservation of the intrinsic data structure TRNMap computes the dissimilarities of the data points based on the graph distances. To compute the graph distances the set of data is replaced by the graph resulted of the TRN algorithm applied on the data set. The edges of the graph are labeled with their Euclidean length and Dijkstra's algorithm [59] is run on the graph, in order to compute the shortest path for each pair of points. The TRNMap algorithm utilises the group of multidimensional scaling mapping algorithms to give the low-dimensional representation of the data set. If the aim of the mapping is the visualisation of the distances of the objects or their representatives, the TRNMap utilises the metric MDS method. On the other hand, if the user is only interested in the ordering relations of the objects, the TRNMap uses non-metric MDS for the low-dimensional representation. As a result it gives compact low-dimensional topology preserving feature maps to explore the hidden structure of data. In the following the TRNMap algorithm is introduced in details.

Given a set of data $\mathbf{X} = \{\mathbf{x}_1, \mathbf{x}_2, \ldots, \mathbf{x}_N\}, \mathbf{x}_i \in \mathbb{R}^D$. The main goal of the algorithm is to give a compact, perspicuous representation of the objects. For this purpose the set of $\mathbf{X}$ is represented in a lower dimensional output space by a new set of the objects $(\mathbf{Y})$, where $\mathbf{Y} = \{\mathbf{y}_1, \mathbf{y}_2, \ldots, \mathbf{y}_n\}, n \leq N, (\mathbf{y}_i \in \mathbb{R}^d, d \ll D)$.

To avoid the influence of the range of the attributes a normalisation procedure is suggested as a preparing step (Step 0). After the normalisation the algorithm creates the topology representing network of the input data set (Step 1). It is achieved by the use of the Topology Representing Network proposed by Martinetz and Shulten [60]. The number of the nodes (representatives) of the TRN is determined by the user. By the use of the TRN, this step ensures the exploration of the correct structure of the data set, and includes a vector quantisation, as well. Contrary to the ε-neighbouring and k-neighbouring algorithm, the graph resulted from applying the TRN algorithm does not depend on the density of the objects or the selected number of the neighbours. If the resulted graph is unconnected, the TRNMap algorithm connects the subgraphs by linking the closest elements (Step 2). Then the pairwise graph distances are calculated between every pair of representatives (Step 3). In the following, the original topology representing network is mapped into a 2-dimensional graph (Step 4). The mapping method utilises the similarity of the data points provided by the previously calculated graph distances. This mapping process can be carried out by the use of metric or non-metric multidimensional scaling, as well. For the expressive visualisation component planes are also created by the D-dimensional representatives (Step 5).

Algorithm 13 Topology Representing Network Map algorithm

Step 0   Normalize the input data set **X**.

Step 1   Create the Topology Representing Network of **X** by the use of the TRN algorithm [60]. Yield $M^{(D)} = (\mathbf{W}, \mathbf{C})$ the resulted graph, let $\mathbf{w}_i \in \mathbf{W}$ be the representatives (codebook vectors) of $M^{(D)}$. If exists an edge between the representatives $\mathbf{w}_i$ and $\mathbf{w}_j$ ($\mathbf{w}_i, \mathbf{w}_j \in \mathbf{W}, i \neq j$), $c_{i,j} = 1$, otherwise $c_{i,j} = 0$.

Step 2   If $M^{(D)}$ is not connected, connect the subgraphs in the following way:

    While   there are unconnected subgraphs ($m_i^{(D)} \subset M^{(D)}, i = 1, 2, \ldots$):

       (a)   Choose a subgraph $m_i^{(D)}$.

       (b)   Let the terminal node $\mathbf{t}_1 \in m_i^{(D)}$ and its closest neighbor
           $\mathbf{t}_2 \notin m_i^{(D)}$ from:
$$\|\mathbf{t}_1 - \mathbf{t}_2\| = min\|\mathbf{w}_j - \mathbf{w}_k\|, \quad \mathbf{t}_1, \mathbf{w}_j \in m_i^{(D)}, \mathbf{t}_2, \mathbf{w}_k \notin m_i^{(D)}$$
       (c)   Set $c_{\mathbf{t}_1, \mathbf{t}_2} = 1$.

    End while

    Yield $M^{*(D)}$ the modified $M^{(D)}$.

Step 3   Calculate the geodesic distances between all $\mathbf{w}_i, \mathbf{w}_j \in M^{*(D)}$.

Step 4   Map the graph $M^{(D)}$ into a 2-dimensional vector space by metric or non-metric MDS based on the graph distances of $M^{*(D)}$.

Step 5   Create component planes for the resulting Topology Representing Network Map based on the values of $\mathbf{w}_i \in M^{(D)}$.

The parameters of the TRNMap algorithm are the same as those of the Topology Representing Networks algorithm. The number of the nodes of the output graph (n) is determined by the user. The bigger the n the more detailed the output map will be. The suggest the choice is $n = 0.2N$, where N yields the number of the original objects. If the number of the input data elements is high, it can result in numerous nodes. In these cases it is practical to decrease the number of the representatives and iteratively run the algorithm to capture the structure more precisely. Values of the other parameters of TRN (λ, the step size ε, and the threshold value of edge's ages T) can be the same as proposed by Martinetz and Schulten [60].

Figure 3.14 shows the 2-dimensional structure of the S curve data set created by the TRNMap method. As TRNMap algorithm utilises geodesic distances to calculate

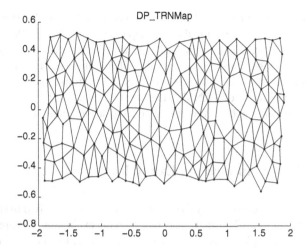

Fig. 3.14 TRNMap visualisation of S curve data set

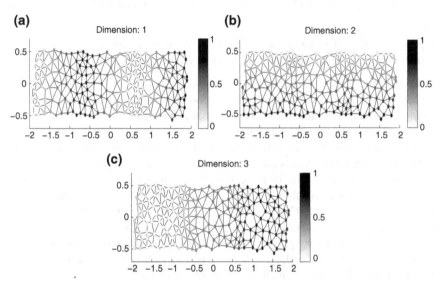

Fig. 3.15 TRNMap component planes of S curve data set. **a** Dimension X. **b** Dimension Y. **c** Dimension Z

the pairwise dissimilarities of the quantised data, this method is able to unfold the real 2-dimensional structure of the S curve data set.

Besides the visualisation of the data structure, the nodes of TRNMap also visualise high-dimensional information by the use of the component plane representation. Component planes of the 3-dimensional S curve data set resulted by the TRNMap are shown in the Fig. 3.15. A component plane displays the value of one component of each node. If the input data set has D attributes, the Topology Representing Network

Map component plane includes D different maps according to the D components. The structure of this map is identical to the map resulted by the TRNMap algorithm, but the nodes are represented in grayscale. White color means the smallest value, black color corresponds to the greatest value of the attribute. By viewing several component maps at the same time it is also easy to see simple correlations between attributes. Because nodes of TRNMap can be seen as possible cluster prototypes, TRNMap can provide the basis for an effective clustering method.

3.6 Analysis and Application Examples

In this section a comparative analysis is given about the previously introduced methods with some examples. The analysis is based on the evaluation of mapping results of the following examples: Swiss roll data set (see Appendix A.6.5), Wine data set (see Appendix A.6.3) and Wisconsin breast cancer data set (see Appendix A.6.4).

The mapping qualities of the algorithms are analysed based on the following two aspects:

- preservation of distance and neighbourhood relations of data, and
- preservation of local and global geometry of data.

In our analysis the distance preservation of the methods is measured by the classical MDS stress function, Sammon stress function and residual variance. The neighbourhood preservation and the local and global mapping qualities are measured by functions of trustworthiness and continuity.

All analysed visualisation methods require the setting of some parameters. In the following the next principle is followed: the identical input parameters of different mapping methods are set in the same way. The common parameters of OVI-NG, GNLP-NG and TRNMap algorithms were in all simulations set as follows: $t_{max} = 200n$, $\varepsilon_i = 0.3$, $\varepsilon_f = 0.05$, $\lambda_i = 0.2n$, $\lambda_f = 0.01$, $T_i = 0.1n$. If the influence of the deletion of edges was not analysed, the value of parameter T_f was set to $T_f = 0.5n$. The auxiliary parameters of the OVI-NG and GNLP-NG algorithms were set as $\alpha_i = 0.3$, $\alpha_f = 0.01$, $\sigma_i = 0.7n$, and $\sigma_f = 0.1$. The value of parameter K in the GNLP-NG method in all cases was set to $K = 2$.

3.6.1 Comparative Analysis of Different Combinations

As in Sect. 3.1 was shown several combinations of vector quantisation, distance calculation and mapping algorithms may serve the low-dimensional representation of high-dimensional data. In this subsection some possible combinations are analysed. For the presentation of the results the Swiss roll data set is chosen, hence it is a typical data set which contains a low-dimensional manifold embedded in a high-dimensional vector space (see Appendix A.6.5).

For comparison of the possible methods different combinations of vector quantisation, distance measure and dimensionality reduction method have been analysed. As vector quantisation the well-known k-means and neural gas algorithms were used. The distances were calculated based on either Euclidean norm (notation: Eu) or graph distance. Although the graph distances can be calculated based on graphs arising from the ε-neighbouring, k-neighbouring or from the Topology Representing Network, only the last two methods (k-neighbouring (knn) and TRN) have been applied, since the ε-neighbouring method is very sensitive to data density. For dimensionality reduction metric and non-metric variants of MDS (notations: mMDS for metric MDS and nmMDS for non-metric MDS) and the Sammon mapping have been applied. As non-metric MDS is performed by an iterative optimisation of the stress function, this method can be stuck in local minima. To avoid this disadvantage the non-metric MDS in all cases was initialised on the result of the metric MDS mapping of the objects. The Sammon mapping was applied without initialisation (Sammon) and with initialisation based on the metric MDS (Sammon_mMDS), where the result of the MDS algorithm serves as the initial projection of the data.

Different combinations require different parameter settings. The number of the representatives in all cases was chosen to be $n = 200$. If k-neighbouring was used for the calculation of geodesic distances, the value of parameter k was chosen to be $k=3$. Parameters of TRN algorithm were tuned according to the rules presented in [60]: $\lambda(t) = \lambda_i(\lambda_f/\lambda_i)^{t/t_{max}}$ and $\varepsilon(t) = \varepsilon_i(\varepsilon_f/\varepsilon_i)^{t/t_{max}}$, where $\lambda_i = 0.2n$, $\lambda_f = 0.01$, $\varepsilon_i = 0.3$, $\varepsilon_f = 0.05$ and $t_{max} = 200n$. Unlike the suggested formula ($T(t) = T_i(T_f/T_i)^{t/t_{max}}$), the threshold of the maximum age of the edges was always kept on $T_i = T_f = 0.1n$. Error values of Sammon stress, metric MDS stress and residual variance were calculated for all combinations. Table 3.1 shows the average error values of running each combination for 10 times.

Table 3.1 shows that mappings based on Euclidean distances are not able to uncover the structure of the data because of the nonlinear embedded manifold. On the other hand, it can be seen that the initialisation of the Sammon mapping with the result of the metric MDS improves the mapping quality. When the distances are calculated based on a graph, the metric MDS results in better mapping quality than the Sammon mapping. Comparing metric and non-metric MDS, we can see that they give similar results. The best mapping results are given by the kmeans+knn+mMDS, kmeans+knn+nmMDS, NG+knn+mMDS, NG+knn+nmMDS, TRN+mMDS and TRN+nmMDS combinations. Comparing all methods it can be seen that the combination of TRN with metric MDS outperforms all other methods.

Naturally, these combinations were tested on other data sets, as well. Tests running on different data sets result in similar error values and conclusions as it has been presented previously. Difference only occurred in the case of analyzing multiclass problems. While the Swiss roll data set contains only a single cluster, real life examples generally contain more groups of objects. In these cases if the mapping method utilises geodesic distances to calculate the low-dimensional presentation of the objects, the graph of the objects or representatives must be connected. Naturally, methods k-neighbouring or TRN not always result in a connected graph. For example, in the case of analysis of iris data set the k-neighbouring method with

Table 3.1 Values of Sammon stress, metric MDS stress and residual variance of different algorithms on the Swiss roll data set

Algorithm	Sammon stress	MDS stress	Res. var.
kmeans+Eu+mMDS	0.05088	0.20743	0.22891
kmeans+Eu+nmMDS	0.05961	0.21156	0.22263
kmeans+Eu+Sammon	0.05084	0.21320	0.24200
kmeans+Eu+Sammon_mMDS	0.04997	0.20931	0.23268
kmeans+knn+mMDS	0.00212	0.00091	0.00326
kmeans+knn+nmMDS	0.00216	0.00091	0.00324
kmeans+knn+Sammon	0.00771	0.00440	0.01575
kmeans+knn+Sammon_mMDS	0.00198	0.00097	0.00348
NG+Eu+mMDS	0.05826	0.04941	0.26781
NG+Eu+nmMDS	0.06659	0.05792	0.26382
NG+Eu+Sammon	0.05758	0.05104	0.27613
NG+Eu+Sammon_mMDS	0.05716	0.05024	0.27169
NG+knn+mMDS	0.00208	0.00086	0.00307
NG+knn+nmMDS	0.00206	0.00087	0.00299
NG+knn+Sammon	0.00398	0.00242	0.00916
NG+knn+Sammon_mMDS	0.00392	0.00237	0.00892
TRN+mMDS	0.00145	0.00063	0.00224
TRN+nmMDS	0.00187	0.00064	0.00221
TRN+Sammon	0.01049	0.00493	0.01586
TRN+Sammon_mMDS	0.00134	0.00068	0.00235

$k = 1, 2, \ldots, 49$ results in two unconnected subgraphs. Consequently, at the creation of a new topology based visualisation algorithm, this observation must be taken into account.

3.6.2 Swiss Roll Data Set

The Swiss roll data set (Fig. 3.16, Appendix A.6.5) is a typical example of the non-linearly embedded manifolds. In this example the number of the representatives in all cases was chosen to be $n = 200$. Linear mapping algorithms, such Principal Component Analysis do not come to the proper result (see Fig. 3.17a), because of the 2-dimensional nonlinear embedding. As can be seen in Fig. 3.17b and c CCA and OVI-NG methods are also unable to uncover the real structure of the data, as they utilise Euclidean distances to calculate the pairwise object dissimilarities.

Figure 3.18 shows the Isotop, CDA and GNLP-NG visualisations of the Swiss roll data set. As CDA and Isotop methods can be based on different vector quantisation methods, both methods were calculated based on the results of k-means and neural gas vector quantisation methods as well. Isotop and CDA require to build a graph to calculate geodesic distances. In these cases graphs were built based on the k-neighbouring approach, and parameter k was set to be $k = 3$. It can be seen that

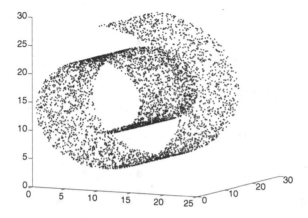

Fig. 3.16 Swiss roll data set

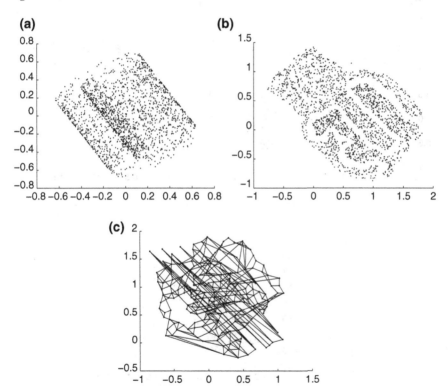

Fig. 3.17 PCA, CCA and OVI-NG projections of Swiss roll data set. **a** 2-dimensional PCA projection. **b** 2-dimensional CCA projection. **c** 2-dimensional OVI-NG projection

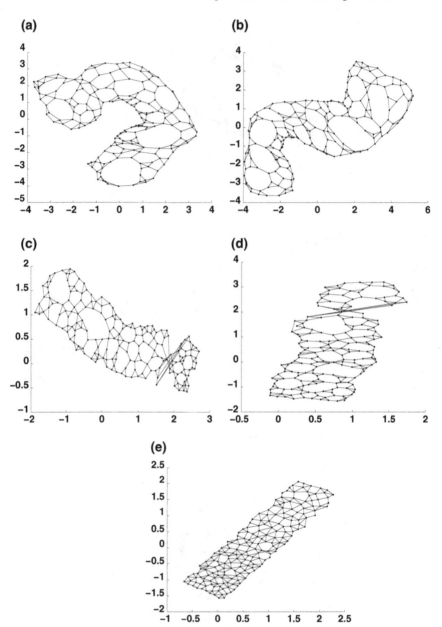

Fig. 3.18 Isotop, CDA and GNLP-NG projections of Swiss roll data set. **a** 2-dimensional Isotop projection with k-means VQ. **b** 2-dimensional Isotop projection with NG VQ. **c** 2-dimensional CDA projection with k-means VQ. **d** 2-dimensional CDA projection with NG VQ. **e** 2-dimensional GNLP-NG projection

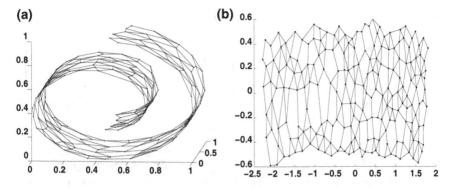

Fig. 3.19 Topology Representing Network and metric MDS based TRNMap visualisation of the Swiss roll data set. **a** TRN. **b** DP_TRNMap

Isotop, CDA and GNLP-NG algorithms can uncover the structure of data in essence, but the Isotop method shows the manifold with some distortions.

In the following let us have a closer look at the results of Topology Representing Network Map algorithm. As TRNMap is based on the creation of the Topology Representing Network, Fig. 3.19a shows the TRN of the Swiss roll data set. In Fig. 3.19b the 2-dimensional metric MDS based TRNMap visualisation of the Swiss roll data set is shown (DP_TRNMap, DP from distance preservation). As the metric MDS and the non-metric MDS based mappings of the resulted TRN in this case give very similar results in the mapped prototypes, the resulted TRNMap visualisations are not distinguishable by human eyes. Thereby Fig. 3.19b can be seen as the result of the non-metric MDS based TRNMap algorithm as well. In Fig. 3.19b it can be seen that the TRNMap methods are able to uncover the embedded 2-dimensional manifold without any distortion.

Visualisation of the Topology Representing Network Map also includes the construction of the component planes. The component planes arising from the metric MDS based TRNMap are shown in Fig. 3.20. The largest value of the attributes of the representatives corresponds to the black and the smallest value to the white dot surrounded by a grey circle. Figure 3.20a shows that alongside the manifold the value of the first attribute (first dimension) initially grows to the highest value, then it decreases to the smallest value, after that it grows, and finally it decreases a little. The second attribute is invariable alongside the manifold, but across the manifold it changes uniformly. The third component starts from the highest value, then it falls to the smallest value, following this it increases to a middle value, and finally it decreases a little.

Table 3.2 shows the error values of distance preservation of different mappings. The notation DP_TRNMap yields the metric MDS based TRNMap algorithm, and the notation NP_TRNMap yields the non-metric MDS based TRNMap algorithm (DP comes from distance preservation and NP from neighbourhood preservation). Table 3.2 shows that GNLP-NG and TRNMap methods outperform the Isotop and

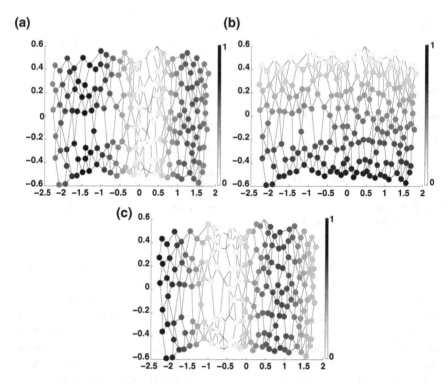

Fig. 3.20 Component planes of the metric MSD based Topology Representing Network Map of the Swiss roll data set. **a** Dimension 1. **b** Dimension 2. **c** Dimension 3

Table 3.2 Values of Sammon stress, metric MDS stress and residual variance of Isotop, CDA, GNLP-NG and TRNMap algorithms on the Swiss roll data set

Algorithm	Sammon stress	Metric MDS stress	Residual variance
kmeans_Isotop	0.54040	0.57870	0.41947
NG_Isotop	0.52286	0.53851	0.15176
kmeans_CDA	0.01252	0.00974	0.01547
NG_CDA	0.01951	0.01478	0.02524
GNLP-NG	0.00103	0.00055	0.00170
DP_TRNMap	0.00096	0.00043	0.00156
NP_TRNMap	0.00095	0.00045	0.00155

CDA methods. Although, GNLP-NG and TRNMap methods show similar performances in distance preservation, the TRNMap methods show somewhat better performances.

Figure 3.21 shows the neighbourhood preservation mapping qualities of the methods to be analysed. It can be seen that different variations of Isotop and CDA show lower performance in neighbourhood preservation than GNLP-NG and TRNMap methods. The continuity and the trustworthiness of GNLP-NG and TRNMap mappings do not show a substantive difference, the qualitative indicators move

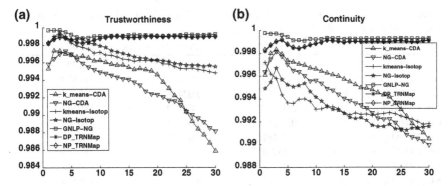

Fig. 3.21 Trustworthiness and continuity as a function of the number of neighbours k, for the Swiss roll data set

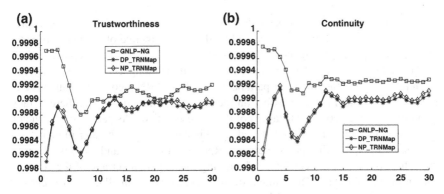

Fig. 3.22 Trustworthiness and continuity of GNLP-NG and TRNMap methods as a function of the number of neighbours k, for the Swiss roll data set

within 0.2 %. Nevertheless, the GNLP-NG method shows better performance in the local area than the TRNMap mappings. For better visibility Fig. 3.22 focuses on the GNLP-NG, metric MDS based TRNMap and non-metric MDS based TRNMap mappings. In this example it has been shown both GNLP-NG and TRNMap methods are able to uncover non-linearly embedded manifolds, the TRNMap methods show good performance both in topology and distance preservation, and the component planes provide useful facilities to unfold the relations among the features.

3.6.3 Wine Data Set

In this subsection a real problems is considered. The wine database (see Appendix A.6.3) contains the chemical analysis of 178 wine, each is characterised by 13 continuous attributes, and there are three classes distinguished.

Table 3.3 Values of Sammon stress, metric MDS stress and residual variance of GNLP-NG and TRNMap algorithms on the Wine data set

Algorithm	Sammon stress	Metric MDS stress	Residual variance
GNLP-NG $T_f = 0.5n$	0.04625	0.03821	0.13926
GNLP-NG $T_f = 0.3n$	0.04982	0.04339	0.15735
GNLP-NG $T_f = 0.05n$	0.02632	0.02420	0.07742
DP_TRNMap $T_f = 0.5n$	0.01427	0.00829	0.03336
DP_TRNMap $T_f = 0.3n$	0.01152	0.00647	0.02483
DP_TRNMap $T_f = 0.05n$	0.01181	0.00595	0.02161
NP_TRNMap $T_f = 0.5n$	0.03754	0.02608	0.07630
NP_TRNMap $T_f = 0.3n$	0.05728	0.04585	0.09243
NP_TRNMap $T_f = 0.05n$	0.03071	0.01984	0.04647

On visualisation results presented in the following the class labels are also presented. The representatives are labeled based on the principle of the majority vote: (1) each data point is assigned to the closest representative; (2) the representatives are labeled with the class label that occurs most often among its assigned data point.

In this example the tuning of parameter T_f of the TRN algorithm is also tested. Parameters T_i and T_f has an effect on the linkage of the prototypes, thereby they also influence the geodesic distances of the representatives. As parameter T_f yields the final threshold of the age of the edges, this parameter has greater influence on the resulted graph. Other parameters of TRN algorithm (λ, ε and t_{max}) were set to the values presented in Sect. 3.6.2. The tuning of the parameter age of the edges is shown in the following. The number of the representatives in all cases was chosen to be $n = 0.2N$, which means 35 nodes in this case.

As parameter T_f has an effect on the creation of the edges of TRN, it influences the results of the GNLP-NG and TRNMap algorithms. Table 3.3 shows the error values of the distance preservation of the GNLP-NG, DP_TRNMap and NP_TRNMap methods. In these simulations parameter T_i was chosen to be $T_i = 0.1n$, and T_f was set to $T_f = 0.5n$, $T_f = 0.3n$ and $T_f = 0.05n$, where n denotes the number of the representatives. It can be seen that the best distance preservation quality is obtained at the parameter setting $T_f = 0.05n$. At parameters $T_f = 0.5n$ and $T_f = 0.3n$ the methods TRNMap based on non-metric MDS and GNLP-NG seem to fall in local minima. It can be caused by their iterative minimizing process. On the contrary, the metric MDS based TRNMap finds the coordinates of the low-dimensional representatives in a single step process by eigenvalues decomposition, and thereby it seems to be a more robust process. It is certified by the good error values of the DP_TRNMap in all three cases.

The effect of the change of the parameter T_f on the visual presentation is shown in Fig. 3.23. It can be seen that the deletion of edges produces smoother graphs.

Based on the previous experimental results, the parameter setting $T_f = 0.05n$ has been chosen for the further analysis. In the following lets have a look at the comparison of the TRNMap methods with the other visualisation methods. Table 3.4

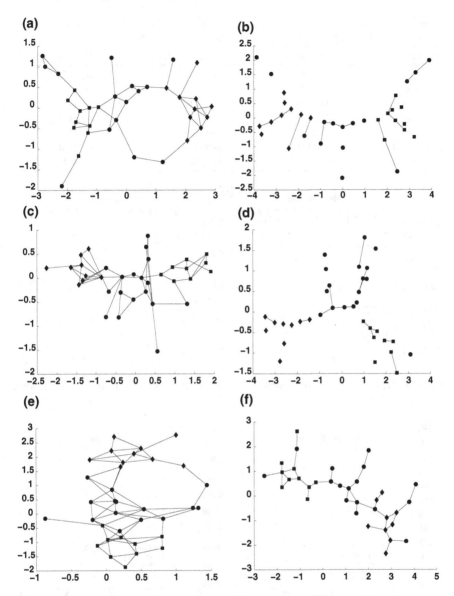

Fig. 3.23 GNLP-NG and TRNMap projections of Wine data set at different settings of parameter T_f. **a** DP_TRNMap $T_f = 0.3n$. **b** DP_TRNMap $T_f = 0.05n$. **c** NP_TRNMap $T_f = 0.3n$. **d** NP_TRNMap $T_f = 0.05n$. **e** GNLP-NG $T_f = 0.3n$. **f** GNLP-NG $T_f = 0.05n$

shows error values of the distance preservation of the methods to be analysed. The parameter k for the k-neighbouring was chosen to be $k = 3$. It can be seen that TRNMap method based on metric MDS mapping shows the best mapping qualities.

Table 3.4 Values of Sammon stress, metric MDS stress and residual variance of Isotop, CDA, GNLP-NG and TRNMap algorithms on the wine data set ($T_f = 0.05$)

Algorithm	Sammon stress	Metric MDS stress	Residual variance
kmeans_Isotop	0.59233	0.59797	0.54959
NG_Isotop	0.60600	0.61030	0.46479
kmeans_CDA	0.93706	0.30726	0.63422
NG_CDA	0.82031	0.27629	0.66418
GNLP-NG	0.02632	0.02420	0.07742
DP_TRNMap	0.01181	0.00595	0.02161
NP_TRNMap	0.03071	0.01984	0.04647

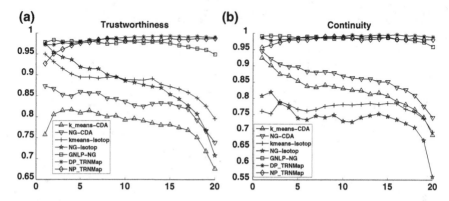

Fig. 3.24 Trustworthiness and continuity as a function of the number of neighbours k, for the wine data set

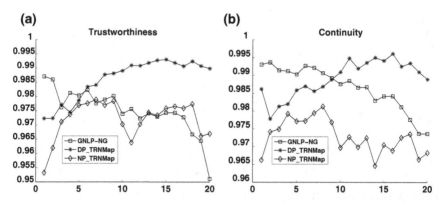

Fig. 3.25 Trustworthiness and continuity of GNLP-NG and TRNMap methods as a function of the number of neighbours k, for the wine data set

Figure 3.24 shows trustworthiness and continuity of different mappings. It can be seen, that GNLP-NG, NP_TRNMap and DP_TRNMap methods give the best performance both in local and in global areas of the objects. For better visibility Fig. 3.25 focuses on the GNLP-NG, metric MDS based TRNMap and non-metric

Table 3.5 The values of the Sammon stress, MDS stress and residual variance of different mapping algorithms on the Wisconsin breast cancer data set ($n = 35$)

Algorithm	Sammon stress	MDS stress	Residual variance
GNLP-NG	0.02996	0.02764	0.09733
DP_TRNMap	0.01726	0.01075	0.04272
NP_TRNMap	0.01822	0.01077	0.03790

MDS based TRNMap mappings. This figure shows that NP_TRNMap method has not found the optimal mapping, because the characteristics of the functions of the NP_TRNMap algorithm differ from the characteristics of functions of DP_TRNMap algorithm. Comparing the GNLP-NG and DP_TRNMap methods we can see that the DP_TRNMap method give better performance at larger k-nn values. Opposite to this the GNLP-NG technique gives better performance at the local reconstruction. (At small k-nn-s the local reconstruction performance of the model is tested, while at larger k-nn-s the global reconstruction is measured.)

3.6.4 Wisconsin Breast Cancer Data Set

Wisconsin breast cancer database is a well-known diagnostic data set for breast cancer (see Appendix A.6.4). This data set contains 9 attributes and class labels for the 683 instances of which 444 are benign and 239 are malignant. It has been shown in the previous examples that the GNLP-NG and TRNMap methods outperform the CDA and Isotop methods both distance and neighbourhood preservation. Thereby, in this example only the qualities of the GNLP-NG and that of the TRNMap methods will be examined. The number of the nodes in this case was reduced to $n = 35$ and $n = 70$. The parameter T_f was chosen to be $T_f = 0.05n$ followed the previously presented method.

To get a compact representation of the data set to be analysed, the number of the neurons was chosen to be $n = 35$ in the beginning. Table 3.5 shows the numerical evaluation of the distance preservation capabilities of the mappings. The efficiency of the TRNMap algorithm in this case is also confirmed by the error values.

TRNMap and GNLP-NG visualisations of the Wisconsin breast cancer data set are shown in Fig. 3.26. The results of the several runs seem to have drawn a fairly wide partition and another compact partition. In these figures the representatives of the benign class are labeled with square markers and the malignant class is yielded with circle markers.

The quality of the neighbourhood preservation of the mappings is shown in Fig. 3.27. Figures illustrate that the MDS-based techniques show better global mapping quality than the GNLP-NG method, but in the local area of the data points the GNLP-NG exceeds the TRNMap methods.

To examine the robustness of the TRNMap methods another number of the representatives has also been tried. In the second case the number of the representatives

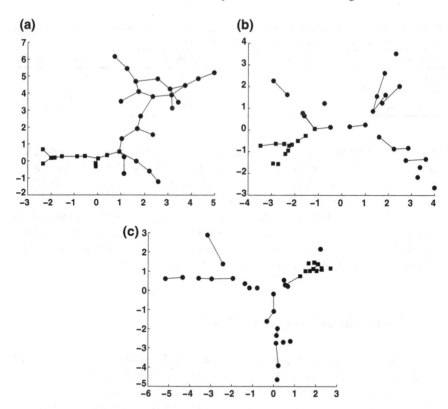

Fig. 3.26 GNLP-NG and TRNMap visualisations of the Wisconsin breast cancer data set. **a** GNLP-NG. **b** DP_TRNMap. **c** NP_TRNMap

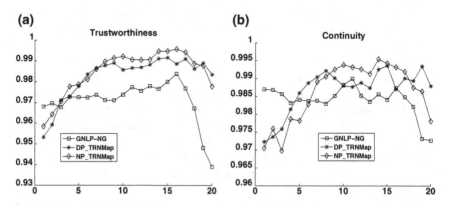

Fig. 3.27 Trustworthiness and continuity as a function of the number of neighbours k, for the Wisconsin breast cancer data set ($n = 35$)

Table 3.6 Values of Sammon stress, MDS stress and residual variance of different mapping algorithms on the Wisconsin breast cancer data set ($n = 70$)

Algorithm	Sammon stress	MDS stress	Residual variance
GNLP-NG	0.06293	0.05859	0.22249
DP_TRNMap	0.01544	0.00908	0.03370
NP_TRNMap	0.02279	0.01253	0.02887

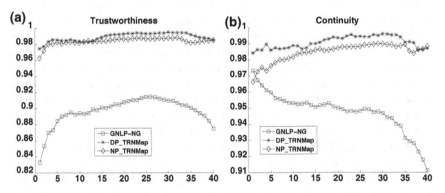

Fig. 3.28 Trustworthiness and continuity as a function of the number of neighbours k, for the Wisconsin breast cancer data set ($n = 70$)

was chosen to be $n = 70$. Table 3.6 and Fig. 3.28 show the numerical evaluations of the methods in this case (other parameters were not changed). Both the error values and the functions show that the GNLP-NG method has fallen again in a local minima. (This incident occurs in many other cases as well.) On the other hand, the TRNMap algorithms in these cases are also robust to the parameter settings.

It is an interesting aspect to compare the error values of the methods in the case of mappings of different data sets (see Tables 3.2, 3.4 and 3.5). Error values of the mappings for the Swiss roll data set are smaller in order of magnitude, than error values in the other two examples. It means, that stress functions of distance preservation are also able to show the presence of such manifolds that can be defined by graphs.

3.7 Summary of Visualisation Algorithms

In this chapter we presented dimensionality reduction methods and examined how these methods are able to visualise hidden structure of data. Robust graph based mapping algorithm have been proposed to visualise the hidden data structure in low-dimensional space. The proposed method is called Topology Representing Network Map (TRNMap), and it provides various mapping solutions. TRNMap combines the main benefits of the Topology Representing Network and the multidimensional scaling. As a result a low-dimensional representation of the data set to be analysed is given, which can reflect both the topology and the metric of the data set. Systematic analysis of the algorithms commonly used for data visualisation and the numerical examples presented in this chapter demonstrate that the resulting map gives a good representation of the topology and the metric of complex data sets, and the component plane representation of TRNMap is a useful tool to explore the hidden relations among the features.

To show the main properties of the presented visualisation methods a detailed analysis has been performed on them. The primary aim of this analysis was to examine the preservation of distances and neighbourhood relations of data. Preservation of neighbourhood relations was analysed both in local and global environments. It has been confirmed that: (1) if low-dimensional manifolds exist in the high-dimensional feature space of the data set algorithms based on geodesic (graph) distances should be preferred over classical Euclidean distance based methods. New scientific statements are: (2) Among the wide range of possible approaches graphs obtained by Topology Representing Networks are the most suitable to approximate this low-dimensional manifold. Thereby, comparing TRNMap, CDA and Isotop algorithms, it can be seen that TRNMap utilises a more efficient calculation of the graph distances than CDA or Isotop. (3) Multidimensional scaling is an effective method to form a low-dimensional map of the TRN based on the calculated graph distances. (4) Component planes of TRNMap provide useful facilities to unfold the relations among the features of the objects to be analysed. (5) Comparing TRNMap and GNLP-NG methods, it can be seen that TRNMap methods are more robust to the initialisation parameters (e.g. such as number of the representatives, or the maximal age of the edges). (6) MDS based techniques can be considered as global reconstruction methods, hence in most cases they give better performances at larger k-nn values. (7) Metric mapping based algorithms (e.g. TRNMap based on metric MDS) directly minimise the stress functions, so their performance is the best in distance perseveration. It is an interesting conclusion that good distance preservation results in good global neighbourhood persevering capabilities. (8) Stress functions of distance preservation are suited to show the presence of such manifolds that can be defined by graphs.

Synthetic and real life examples have shown that Topology Representing Network Map utilises the advantages of several dimensionality reduction methods so that it is able to give a compact representation of low-dimensional manifolds nonlinearly embedded in the high-dimensional feature space.

References

1. Tukey, J.: Exploratory Data Analysis. Addison-Wesley, New York (1977)
2. McQueen, J.: Some methods for classification and analysis of multivariate observations. Proceedings of fifth Berkeley symposium on mathematical statistics and probability, pp. 281–297 (1967)

3. Martinetz, T.M., Shulten, K.J.: A neural-gas network learns topologies. In: Kohonen, T., Mäk-isara, K., Simula, O., Kangas, J. (eds.) Artificial Neural Networks, pp. 397–402. Elsevier Science Publishers B.V, North-Holland (1991)

4. Johannes, M., Brase, J.C., Fröhlich, H., Gade, S., Gehrmann, M., Fälth, M., Sültmann, H., Beißbarth, T.: Integration of pathway knowledge into a reweighted recursive feature elimination approach for risk stratification of cancer patients. Bioinformatics 26(17), 2136–2144 (2010)

5. Lai, C., Reinders, M.J.T., Wessels, L.: Random subspace method for multivariate feature selection. Pattern Recognit. Lett. 27(10), 1067–1076 (2006)

6. Nguyen, M.H., de la Torre, F.: Optimal feature selection for support vector machines. Pattern Recognit. 43(3), 584–591 (2010)

7. Rong, J., Li, G., Chen, Y.P.P.: Acoustic feature selection for automatic emotion recognition from speech. Inf. Process. Manag. 45(3), 315–328 (2009)

8. Tsang, I.W., Kocsor, A., Kwok, J.T.: Efficient kernel feature extraction for massive data sets. Proceedings of the 12th ACM SIGKDD international conference on Knowledge discovery and data mining, pp. 724–729 (2006)

9. Wang, J., Zhang, B., Wang, S., Qi, M., Kong, J.: An adaptively weighted sub-pattern locality preserving projection for face recognition. J. Netw. Comput. Appl. 332(3), 323–332 (2010)

10. Blum, A.L., Langley, P.: Selection of relevant features and examples in machine learning. Artif. Intell. 97(12), 245271 (1997)

11. Guyon, I., Elisseeff, A.: An introduction to variable and feature selection. J. Mach. Learn. Res. 3, 1157–1182 (2003)

12. Jain, A., Zongker, D.: Feature selection: Evaluation, application, and small sample performance. IEEE Trans. Pattern Anal. Mach. Intell. 192, 153–158 (1997)

13. Weston, J., et al.: Feature selection for SVMs. In: Leen, T.K., Dietterich, T.G., Tresp, V. (eds.) Advances in Neural Information Processing Systems, vol. 13, pp. 668–674. The MIT Press, Cambride (2001)

14. Narendra, P., Fukunaga, K.: A branch and bound algorithm for feature subset selection. IEEE Trans. Comput. C–26(9), 917–922 (1977)

15. Pudil, P., Novovičová, J., Kittler, J.: Floating search methods in feature selection. Pattern Recognit. Lett. 15(1), 1119–1125 (1994)

16. Hotelling, H.: Analysis of a complex of statistical variables into principal components. J. Educ. Psychol. 24, 417–441 (1933)

17. Jolliffe, T.: Principal Component Analysis. Springer, New York (1996)

18. Sammon, J.W.: A non-linear mapping for data structure analysis. IEEE Trans. Comput. 18(5), 401–409 (1969)

19. Tenenbaum, J.B., Silva, V., Langford, J.C.: A global geometric framework for nonlinear dimensionality reduction. Science 290, 2319–2323 (2000)

20. Comon, P.: Independent component analysis: a new concept? Signal Process. 36(3), 287–317 (1994)

21. Fisher, R.A.: The use of multiple measurements in taxonomic problems. Ann. Eugen. 7, 179–188 (1936)

22. Kohonen, T.: Self-organized formation of topologically correct feature maps. Biol. Cybern. 43, 59–69 (1982)

23. Kohonen, T.: Self-Organizing maps, 3rd edn. Springer, New York (2001)

24. Roweis, S.T., Saul, L.K.: Nonlinear dimensionality reduction by locally linear embedding. Science 290, 2323–2326 (2000)

25. Saul, L.K., Roweis, S.T.: Think globally, fit locally: unsupervised learning of low dimensional manifolds. J. Mach. Learn. Res. 4, 119–155 (2003)

26. Belkin, M., Niyogi, P.: Laplacian eigenmaps for dimensionality reduction and data representation. Neural Comput. 15(6), 1373–1396 (2003)

27. Borg, I.: Modern multidimensional scaling: theory and applications. Springer, New York (1977)

28. Kruskal, J.B., Carroll, J.D.: Geometrical models and badness-of-fit functions. In: Krishnaiah, R. (ed.) Multivariate Analysis II, vol. 2, pp. 639–671. Academic Press Pachuri, New York (1969)

29. Kaski, S., Nikkilä, J., Oja, M., Venna, J., Törönen, J., Castrén, E.: Trustworthiness and metrics in visualizing similarity of gene expression. BMC Bioinformatics **4**, 48 (2003)
30. Venna, J., Kaski, S.: Local multidimensional scaling with controlled tradeoff between trustworthiness and continuity, In: Proceedings of the workshop on self-organizing maps, pp. 695–702 (2005)
31. Venna, J., Kaski, S.: Local multidimensional scaling. Neural Netw. **19**(6), 889–899 (2006)
32. Kiviluoto, K.: Topology preservation in self-organizing maps. Proceedings of IEEE international conference on neural networks, pp. 294–299 (1996)
33. Bauer, H.U., Pawelzik, K.R.: Quantifying the neighborhood preservation of selforganizing feature maps. IEEE Trans. Neural Netw. **3**(4), 570–579 (1992)
34. Duda, R.O., Hart, P.E., Stork, D.: Pattern classification. Wiley, New York (2000)
35. Mika, S., Schölkopf, B., Smola, A.J., Müller, K.-R., Scholz, M., Rätsch, G.: Kernel PCA and de-noising in feature spaces. In: Advances in neural information processing systems, vol. 11, Cambridge, USA (1999)
36. Schölkopf, B., Smola, A.J., Müller, K.-R.: Nonlinear component analysis as a kernel eigenvalue problem. Neural Comput. **10**(5), 1299–1319 (1998)
37. Mao, J., Jain, A.K.: Artifical neural networks for feature extraction and multivariate data projection. IEEE Trans. Neural Netw 6(2), 629–637 (1995)
38. Pal, N.R., Eluri, V.K.: Two efficient connectionist schemes for structure preserving dimensionality reduction. IEEE Trans. Neural Netw. **9**, 1143–1153 (1998)
39. Young, G., Householder, A.S.: Discussion of a set of points in terms of their mutual distances. Psychometrika **3**(1), 19–22 (1938)
40. Naud, A.: Neural and statistical methods for the visualization of multidimensional data. Technical Science Katedra Metod Komputerowych Uniwersytet Mikoaja Kopernika w Toruniu (2001)
41. Kruskal, J.B.: Multidimensional scaling by optimizing goodness-of-fit to a nonmetric hypothesis. Psychometrika **29**, 1–29 (1964)
42. He, X., Niyogi, P.: Locality preserving projections. In: Lawrence, K., Saul, Weiss, Y., Bottou, L., (eds.) Advances in Neural Information Processing Systems 17. Proceedings of the 2004 Conference, MIT Press, vol. 16, p. 37 (2004) http://mitpress.mit.edu/books/advances-neural-information-processing-systems-17
43. UC Irvine Machine Learning Repository www.ics.uci.edu/ mlearn/ Cited 15 Oct 2012
44. Haykin, S.: Neural Networks: A Comprehensive Foundation. Prentice Hall, Upper Saddle River (1999)
45. Ultsch, A.: Self-organization neural networks for visualization and classification. In: Opitz, O., Lausen, B., Klar, R. (eds.) Information and Classification, pp. 307–313. Springer, Berlin (1993)
46. Blackmore, J., Miikkulainen, R.: Incremental grid growing: encoding high-dimensional structure into a two-dimensional feature map. In: Proceedong on IEEE international conference on neural networks, vol. 1, pp. 450–455 (1993)
47. Merkl, D., He, S.H., Dittenbach, M., Rauber, A.: Adaptive hierarchical incremental grid growing: an architecture for high-dimensional data visualization. In: Proceeding of the workshop on SOM, Advances in SOM, pp. 293–298 (2003)
48. Lee, J.A., Lendasse, A., Donckers, N., Verleysen, M.: A robust nonlinear projection method. Proceedings of ESANN'2000, 8th european symposium on artificial, neural networks, pp. 13–20 (2000)
49. Estévez, P.A., Figueroa, C.J.: Online data visualization using the neural gas network. Neural Netw. **19**, 923–934 (2006)
50. Estévez, P.A., Chong, A.M., Held, C.M., Perez, C.A.: Nonlinear projection using geodesic distances and the neural gas network. Lect. Notes Comput. Sci. **4131**, 464–473 (2006)
51. Wu, Y., Chan, K.L.: An extended isomap algorithm for learning multi-class manifold. Proceeding of IEEE international conference on machine learning and, cybernetics (ICMLC2004), vol. 6, pp. 3429–3433 (2004)

52. Lee, J.A., Verleysen, M.: Nonlinear projection with the isotop method. Proceedings of ICANN'2002, international conference on artificial, neural networks, pp. 933–938 (2002)
53. Lee, J.A., Archambeau, C., Verleysen, M.: Locally linear embedding versus isotop. In: ESANN'2003 proceedings: European symposium on artificial neural networks Bruges (Belgium), pp. 527–534 (2003)
54. Demartines, P., Herault, J.: Curvilinear component analysis: a self-organizing neural network for nonlinear mapping of data sets. IEEE Trans. Neural Netw. **8**, 148–154 (1997)
55. Lee, J.A., Lendasse, A., Verleysen, M.: Curvilinear distance analysis versus isomap. Proceedings of ESANN'2002, 10th European symposium on artificial, neural networks, pp. 185–192 (2000)
56. Lee, J.A., Lendasse, A., Verleysen, M.: Nonlinear projection with curvilinear distances: isomap versus curvilinear distance analysis. Neurocomputing **57**, 49–76 (2004)
57. Vathy-Fogarassy, A., Kiss, A., Abonyi, J.: Topology representing network map—a new tool for visualization of high-dimensional data. Trans. Comput. Sci. I. **4750**, 61–84 (2008) (Springer)
58. Vathy-Fogarassy, A., Abonyi, J.: Local and global mappings of topology representing networks. Inf. Sci. **179**, 3791–3803 (2009)
59. Dijkstra, E.W.: A note on two problems in connection with graphs. Numer. Math. **1**, 269–271 (1959)
60. Martinetz, T.M., Shulten, K.J.: Topology representing networks. Neural Netw. **7**(3), 507–522 (1994)

Appendix

A.1 Constructing a Minimum Spanning Tree

A.1.1 Prim's Algorithm

Prim's algorithm is a greedy algorithm to construct the minimal spanning tree of a graph. The algorithm was independently developed in 1930 by Jarník [1] and later by Prim in 1957 [2]. Therefore the algorithm is sometimes called as Jarník algorithm as well. This greedy algorithm starts with one node in the 'tree' and iteratively step by step adds the edge with the lowest cost to the tree. The formal algorithm is given in Algorithm 14.

Algorithm 14 Prim's algorithm

Given a non-empty connected weighted graph $G = (V, E)$, where V yields the set of the vertices and E yields the set of the edges.

Step 1   Select a node ($x \in V$) arbitrary from the vertices. This node will be the root in the tree. Set $V_{new} = \{x\}$, $E_{new} = \{\}$.

Step 2   Choose the edge with the lowest cost from the set of the edges $e(u, v)$ such that $u \in E_{new}$ and $v \notin E_{new}$. If there are multiple edges with the same weight connecting to the tree constructed so far, any of them may be selected. Set $E_{new} = E_{new} \cup \{v\}$ and $V_{new} = V_{new} \cup \{e(u, v)\}$.

Step 3   If $V_{new} \neq V$ go back to Step 2.

A.1.2 Kruskal's Algorithm

Kruskal's algorithm [3] is an another greedy algorithm to construct the minimal spanning tree. This algorithm starts with a forest (initially each node of the graph represents a tree) and iteratively adds the edge with the lowest cost to the forest connecting trees such a way, that circles in the forest are not enabled. The detailed algorithm is given in Algorithm 15.

Á. Vathy-Fogarassy and J. Abonyi, *Graph-Based Clustering*
and Data Visualization Algorithms, SpringerBriefs in Computer Science,
DOI: 10.1007/978-1-4471-5158-6, © János Abonyi 2013

Algorithm 15 Kruskal's algorithm

Given a non-empty connected weighted graph $G = (V, E)$, where V yields the set of the vertices and E yields the set of the edges.

Step 1   Create a forest F in such a way that each vertex (V) of the graph G denotes a separate tree. Let $S = E$.

Step 2   Select an edge e from S with the minimum weight. Set $S = S - \{e\}$.

Step 3   If the selected edge e connects two separated trees, then add it to the forest.

Step 4   If F is not yet a spanning tree and $S \neq \{\}$ go back to Step 2.

A.2 Solutions of the Shortest Path Problem

A.2.0.1 Dijkstra's Algorithm

Dijkstra's algorithm calculates the shortest path from a selected vertex to every other vertex in a weighted graph where the weights of the edges non-negative numbers. Similar to Prim's and Kruskal's algorithms it is a greedy algorithm, too. It starts from a selected node s and iteratively adds the closest node to a so far visited set of nodes. The whole algorithm is described in Algorithm 16.

At the end of the algorithm the improved tentative distances of the nodes yields their distances to vertex s.

Algorithm 16 Dijkstra's algorithm

Given a non-empty connected weighted graph $G = (V, E)$, where V yields the set of the vertices and E yields the set of the edges. Let $s \in V$ the initial vertex from which we want to determine the shortest distances to the other vertices.

Step 0   Initialization: Denote $P_{visited}$ the set of the nodes visited so far and $P_{unvisited}$ the set of the nodes unvisited so far. Let $P_{visited} = \{s\}$, $P_{unvisited} = V - \{s\}$. Yield the $v_{current}$ the current node and set $v_{current} = s$. Assign a tentative distance to each node as follows: set it to 0 for node s ($dist_s = 0$) and to infinity for all other nodes.

Step 1   Consider all unvisited direct topological neighbors of the current node, and calculate their tentative distances to the initial vertex as follows:

$$dist_{s,e_i} = dist_{s,v_{current}} + w_{v_{current},e_i}, \tag{A.1}$$

where $e_i \in P_{unvisited}$ is a direct topological neighbor of $v_{current}$, and $w_{v_{current},e_i}$ yields the weight of the edge between nodes $v_{current}$ and e_i.

If the recalculated distance for e_i is less than the previously calculated distance for e_i, than record it as its new tentative distance to s.

Step 2   Mark the current vertex as visited by setting $P_{visited} = P_{visited} \cup \{v_{current}\}$, and remove it from the unvisited set by setting $P_{unvisited} = P_{unvisited} - \{v_{current}\}$.

Step 3   If $P_{unvisited} = \{\}$ or if the smallest tentative distance among the nodes in the unvisited set is infinity then stop. Else select the node with the smallest tentative distance to s from the set of the unvisited nodes, set this node as the new current node and go back to Step 1.

A.2.1 Floyd-Warshall Algorithm

Given a weighted graph $G = (V, E)$. The Floyd-Warshall algorithm (or Floyd's algorithm) computes the shortest path between all pairs of the vertices of G. The algorithm operates on a $n \times n$ matrix representing the costs of edges between vertices, where n is the number of the vertices in G ($\|V\| = n$). The elements of the matrix are initialized and step by step updated as it is described in Algorithm 17.

Algorithm 17 Floyd-Warshall algorithm

Given a weighted graph $G = (V, E)$, where V yields the set of the vertices and E yields the set of the edges. Let $n = \|V\|$ the number of the vertices. Denote D the matrix with dimension $n \times n$ that stores the lenghts of the paths between the vertices. After running the algorithm elements of D will contain the lengths of the shortest paths between all pairs of edges G.

Step 1 Initialization of matrix D: If there is an edge between vertex i ($v_i \in V$) and vertex j ($v_j \in V$) in the graph, the cost of this edge is placed in position (i, j) and (j, i) of the matrix. If there is no edge directly linking two vertices, an infinite (or a very large) value is placed in the positions (i, j) and (j, i) of the matrix. For all $v_i \in V$, $i = 1, 2, \ldots, n$ the elements (i, i) of the matrix is set to zero.

Step 2 Recalculating the elements of D.

for $(k = 1; k <= n; k++)$
  for $(i = 1; i <= n; i++)$
    for $(j = 1; j <= n; j++)$
      if $D_{i,j} > (D_{i,k} + D_{k,j})$ then set $D_{i,j} = D_{i,k} + D_{k,j}$

A.3 Hierarchical Clustering

Hierarchical clustering algorithms may be divided into two main groups: (i) agglomerative methods and (ii) divisive methods. The *agglomerative hierarchical methods* start with N clusters, where each cluster contains only an object, and they recursively merge the two most similar groups into a single cluster. At the end of the process the objects will form only a single cluster. The *divisive hierarchical methods* begin with all objects in a single cluster and perform splitting until all objects form a discrete partition.

All hierarchical clustering methods work on similarity or distance matrices. The agglomerative algorithms merge step by step those clusters that are the most similar and the divisive methods split those clusters that are most dissimilar. The similarity or distance matrices are updated step by step trough the iteration process. Although, the similarity or dissimilarity matrices are generally obtained from the Euclidian distances of pairs of objects, the pairwise similarities of the clusters are definable on a numerous other ways.

The *agglomerative hierarchical methods* utilize most commonly the following approaches to determine the distances between the clusters: (i) single linkage method, (ii) complete linkage method and (iii) average linkage method. The *single linkage method* [4] is also known as the nearest neighbor technique. Using this similarity measure the agglomerative hierarchical algorithms join together the two clusters whose two closest members have the smallest distance. The single linkage clustering methods are also often utilized in the graph theoretical algorithms, however these methods suffer from the chaining effect [5]. The *complete linkage methods* [6] (also known as the furthest neighbor methods) calculate the pairwise cluster similarities based on the furthest elements of the clusters. These methods merge the two clusters with the smallest maximum pairwise distance in each step. Algorithms based on complete linkage methods produce tightly bound or compact clusters [7]. The *average linkage methods* [8] consider the distance between two clusters to be equal to the average distance from any member of one cluster to any member of the other cluster. These methods merge those clusters where this average distance is the minimal. Naturally, there are other methods to determine the merging condition, e.g. the *Ward method*, in which the merging of two clusters is based on the size of an error sum of squares criterion [9].

The *divisive hierarchical methods* are computationally demanding. If the number of the objects to be clustered is N, there are $2^{N-1} - 1$ possible divisions to form the next stage of the clustering procedure. The division criterion may be based on a single variable (monothetic divisive methods) or the split can also be decided by the use of all the variables simultaneously (polythetic divisive methods).

The nested grouping of objects and the similarity levels are usually displayed in a *dendrogram*. The dendrogram is a tree-like diagram, in which the nodes represent the clusters and the lengths of the stems represent the distances of the clusters to be merged or split. Figure A.1 shows a typical dendrogram representation. It can be seen, in the case of the application of an agglomerative hierarchical method, objects a and b will be merged first, then the objects c and d will coalesce into a group, following this, the algorithm merges the clusters containing the objects $\{e\}$ and $\{c, d\}$. Finally, all objects would belong to a single cluster.

One of the main advantages of the hierarchical algorithms is that the number of the clusters need not be specified a priori. There are several possibilities to chose the proper result from the nested series of the clusters. On the one hand, it is possible to stop the running of a hierarchical algorithm when the distance between the nearest

Fig. A.1 Dendrogram

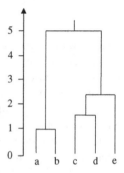

clusters exceeds a predetermined threshold, or on the other hand the dendrogram also offers a useful tool in the selection of the optimal result. The shape of the dendrogram informally suggests the number of the clusters and hereby the optimal clustering result.

Hierarchical clustering approaches are in close ties with graph based clustering. One of the best-known graph-theoretic divisive clustering algorithm (Zahn's algorithm [10]) is based on the construction of the minimal spanning tree. This algorithm step by step eliminates the 'inconsistent' edges from the graph and hereby results in a series of subgraphs.

A.4 Visual Assessment of Cluster Tendency

Visual Assessment of Cluster Tendency (VAT) is an effective and interesting visualization method to reveal the number and the structure of clusters. The method of VAT was proposed in [11], and its variants in [12, 13]. Its aim is similar to one of cluster validity indices, but it tries to avoid the 'massive aggregation of information' by scalar validity measures. Instead of a scalar value or a series of scalar values by different number of clusters, an $N \times N$ intensity image is proposed by Hathaway and Bezdek. It displays the reordered form of the dissimilarity data $\mathbf{D} = [d(\mathbf{x}_i, \mathbf{x}_j)]_{N \times N}$, where $d(\mathbf{x}_i, \mathbf{x}_j)$ is the dissimilarity (not necessarily distance) of the ith and jth samples. The method consists of two steps.

- **Step 1** reorder the dissimilarity data and get $\tilde{\mathbf{D}}$, in which the adjacent points are members of a possible cluster;
- **Step 2** display the dissimilarity image based on $\tilde{\mathbf{D}}$, where the gray level of a pixel is in connection with the dissimilarity of the actual pair of points.

The key step of this procedure is the *reordering* of $\mathbf{D}$. For that purpose, Bezdek used Prim's algorithm [2] (see Appendix A.1.1) for finding a minimal spanning tree. The undirected, fully connected and weighted graph analysed here contains the data points or samples as nodes (vertices) and the edge lengths or weights of the edges are the values in $\mathbf{D}$, the pairwise distances between the samples. There are two differences between Prim's algorithm and VAT: (1) VAT does not need the minimal spanning tree itself (however, it determine also the edges but does not store them), just the order in which the vertices (samples or objects $\mathbf{x}_i$) are added to the tree; and (2) it applies special initialization. Minimal spanning tree contains all of the vertices of the fully connected, weighted graph of the samples, therefore any points can be selected as initial vertex. However, to help ensure the best chance of display success, Bezdek proposed a special initialization: the initial vertex is any of the two samples that are the farthest from each other in the data set ($\mathbf{x}_i$, where i is the row or column index of $\max(\mathbf{D})$). The first row and column of $\tilde{\mathbf{D}}$ will be ith row and column in $\mathbf{D}$. After the initialization, the two methods are exactly the same. Namely, $\mathbf{D}$ is reordered so that the second row and column correspond to the sample closest to the first sample, the third row and column correspond to the sample closest either one of the first two samples, and so on.

This procedure is similar to the single-linkage algorithm that corresponds to the Kruskal's minimal spanning tree algorithm [3] (see Appendix A.1.2) and is basically the greedy approach to find a minimal spanning tree. By hierarchical clustering algorithms (such as single-linkage, complete-linkage or average-linkage methods), the results are displayed as a dendrogram, which is a nested structure of clusters. (Hierarchical clustering methods are not described here, the interested reader can refer e.g. [8]). Bezdek et al. followed another way and they displayed the results as an intensity image $I(\tilde{\mathbf{D}})$ with the size of $N \times N$. The approach was presented in [13] as follows. Let $G = \{g_m, \ldots, g_M\}$ be the set of gray levels used for image displays. In the following, $G = \{0, \ldots, 255\}$, so $g_m = 0$ (black) and $g_M = 255$ (white). Calculate

$$(I(\tilde{\mathbf{D}}))_{i,j} = \tilde{\mathbf{D}}_{i,j} \left(\frac{g_M}{\max(\tilde{\mathbf{D}})} \right). \qquad (A.2)$$

Convert $(I(\tilde{\mathbf{D}}))_{i,j}$ to its nearest integer. These values will be the intensity displayed for pixel (i, j) of $I(\tilde{\mathbf{D}})$. In this form of display, 'white' corresponds to the maximal distance between the data (and always will be two white pixels), and the darker the pixel the closer the two data are. (For large data sets, the image can easily exceed the resolution of the display. To solve that problem, Huband, Bezdek and Hathaway have been proposed variations of VAT [13]). This image contains information about cluster tendency. Dark blocks along the diagonal indicate possible clusters, and if the image exhibits many variations in gray levels with faint of indistinct dark blocks along the diagonal, then the data set "[...] does not contain distinct clusters; or the clustering scheme implicitly imbedded in the reordering strategy fails to detect the clusters (there are cluster types for which single-linkage fails famously [...])."

Figure A.2 gives a small example for the VAT representation. In this example the number of the objects is 40, thereby VAT represents the data dissimilarities in a square image with 40×40 pixels. The figure shows how the well-separated cluster structure is indicated by dark diagonal blocks in the intensity image. Although VAT becomes intractable for large data sets, the bigVAT [13] as a modification of VAT allows the visualization for larger data sets, too.

One of the main advantages of hierarchical clusterings is that they are able to detect non-convex clusters. It is e.g. an 'S'-like cluster in two dimensions; and it can be the case that two data points, which clearly belong to the same cluster, are relatively far from each other. In this case, the dendrogram generated by single-linkage clearly indicates the distinct clusters, but there will be no dark block in the intensity image by VAT. Certainly, single-linkage does have the drawbacks, e.g. it suffers from chaining effect, but a question naturally comes up: how much plus information can be given by VAT? It is because it roughly does a hierarchical clustering, but the result is not displayed as a dendrogram but based on the pairwise distance of data samples, and it works well only if the data in the same cluster are relatively close to each other

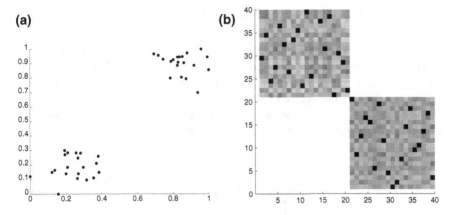

Fig. A.2 The VAT representation. **a** The original data set. **b** VAT

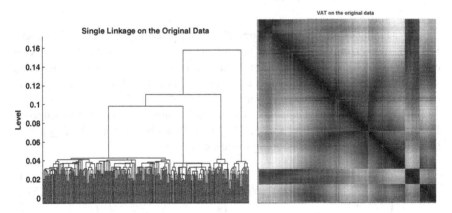

Fig. A.3 Result of the (*left*) single-linkage algorithm and (*right*) VAT on synthetic data

based on the *original distance norm*. (This problem arises not only by clusters with non-convex shape, but very elongated ellipsoids as well.) Therefore, one advantage of hierarchical clustering is lost.

In Fig. A.3 results of the single-linkage algorithm and VAT can be seen on the synthetic data. The clusters are well-separated but non-convex, and single-linkage clearly identifies them as can be seen from the dendrogram. However, the VAT image is not as clear as the dendrogram in this case because there are data in the 'S' shaped cluster that are far from each other based on the Euclidean distance norm (see the top and left corner of the image).

A.5 Gath-Geva Clustering Algorithm

Fuzzy clustering methods assign degrees of membership in several clusters to each input pattern. The resulted *fuzzy partition matrix* (U) describes the relationship of the objects and the clusters. The fuzzy partition matrix $U = [\mu_{i,k}]$ is a $c \times N$ matrix, where $\mu_{i,k}$ denotes the degree of the membership of x_k in cluster C_i, so the i-th row of U contains values of the membership function of the i-th fuzzy subset of X. Conditions of the fuzzy partition matrix are given:

$$\mu_{i,k} \in [0, 1], \quad 1 \le i \le c, \quad 1 \le k \le N, \tag{A.3}$$

$$\sum_{i=1}^{c} \mu_{i,k} = 1, \quad 1 \le k \le N, \tag{A.4}$$

$$0 < \sum_{k=1}^{N} \mu_{i,k} < N, \quad 1 \le i \le c. \tag{A.5}$$

The meaning of the relations written above is that the degree of the membership is a real number in [0,1] (A.3); the sum of the membership values of an object is exactly one (A.4); each cluster must contain at least one object with membership value larger than zero, and the sum of the degrees of the membership values can not exceed the number of elements considered (A.5).

Based on the previous statements the fuzzy partitioning space can be formulated as follows: Let $X = \{x_1, x_2, \ldots, x_N\}$ be a finite set of the observed data, let $2 \le c \le N$ be an integer. The fuzzy partitioning space for X is the set

$$M_{fc} = \left\{ U \in \mathbb{R}^{c \times n} | \mu_{i,k} \in [0, 1], \forall i, k; \sum_{i=1}^{c} \mu_{i,k} = 1, \forall k; 0 < \sum_{k=1}^{N} \mu_{i,k} < N, \forall i \right\}$$

$$\tag{A.6}$$

A common limitation of partitional clustering algorithms based on a fixed distance norm, like k-means or fuzzy c-means clustering is, that they induce a fixed topological structure and force the objective function to prefer clusters of spherical shape even if it is not present. Generally, different cluster shapes (orientations, volumes) are required for the different clusters, but there is no guideline as to how to choose them a priori.

Mixture-resolving methods (e.g. Gustafson-Kessel, Gath-Geva) assume that the objects to be clustered are drawn from one of several distributions (usually Gaussian), and hereby different clusters may form different shapes and sizes. The main task and at the same time the main difficulty of these methods is to estimate the parameters of all these distributions. These algorithms apply several norm-inducing matrices to estimate the data covariance in each cluster. Most of the mixture-resolving methods assume that the individual components of the mixture density are Gaussians, and in this case the parameters of the individual Gaussians are to be estimated by the procedure.

Traditional approaches to this problem involve obtaining (iteratively) a maximum likelihood estimate of the parameter vectors of the component densities [8]. More recently, the Expectation Maximization (EM) algorithm (a general purpose maximum likelihood algorithm [14] for missing-data problems) has been applied to the problem of parameter estimation. The Gustafson-Kessel (GK) [15] and the Gaussian mixture based fuzzy maximum likelihood estimation (Gath-Geva algorithm (GG) [16]) algorithms are also based on an adaptive distance norm, and they are able to estimate the underlying distribution of the objects. Hereby, these algorithms are able to disclose clusters with different orientation and volume. The Gath-Geva (GG) algorithm can be seen as a further development of the Gustafson-Kessel algorithm. In Gath-Geva algorithm the cluster sizes are not restricted like in Gustafson-Kessel method, and the cluster densities are also taken into consideration. Unfortunately, the GG algorithm is very sensitive to initialization, hence often it can not be directly applied to the data.

As we have seen, the exploration of the shapes of the clusters is an essential task. The shape of the clusters can be determined by the distance norm. The typical distance norm between the object $\mathbf{x}_k$ and the cluster center $\mathbf{v}_i$ is represented as:

$$D_{i,k}^2 = \|\mathbf{x}_k - \mathbf{v}_i\|_{\mathbf{A}}^2 = (\mathbf{x}_k - \mathbf{v}_i)^T \mathbf{A} (\mathbf{x}_k - \mathbf{v}_i), \qquad (A.7)$$

where $\mathbf{A}$ is a symmetric and positive definite matrix. Different distance norms can be induced by the choice of the matrix $\mathbf{A}$. The Euclidean distance arises with the choice of $\mathbf{A} = \mathbf{I}$ where $\mathbf{I}$ is an identity matrix. The Mahalanobis normis induced when $\mathbf{A} = \mathbf{F}^{-1}$ where $\mathbf{F}$ is the covariance matrix of the objects. It can be seen that both the Euclidean and the Mahalanobis distances are based on fixed distance norms. The Euclidean norm based methods find only hyperspherical clusters, and the Mahalanobis norm based methods find only hyperellipsoidal ones (see Fig. A.4) even

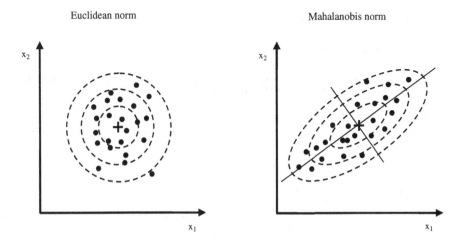

Fig. A.4 Different distance norms in fuzzy clustering

if those shapes of clusters are not present in the data set. The norm-inducing matrix of the cluster prototypes can be adapted by using estimates of the data covariance, and can be used to estimate the statistical dependence of the data in each cluster. The Gustafson-Kessel algorithm (GK) [15] and the Gaussian mixture based fuzzy maximum likelihood estimation algorithm (Gath-Geva algorithm (GG) [16]) are based on such an adaptive distance measure, they can adapt the distance norm to the underlying distribution of the data which is reflected in the different sizes of the clusters, hence they are able to detect clusters with different orientation and volume.

Algorithm 18 Gath-Geva algorithm

Given a set of data $\mathbf{X}$, specify the number of the clusters c, choose a weighting exponent $m > 1$ and a termination tolerance $\varepsilon > 0$. Initialize the partition matrix $\mathbf{U}^{(0)}$.

Repeat   for $t = 1, 2, \ldots$

Step 1   Calculate the cluster centers: $\mathbf{v}_i^{(t)} = \dfrac{\sum_{k=1}^{N} (\mu_{i,k}^{(t-1)})^m \mathbf{x}_k}{\sum_{k=1}^{N} (\mu_{i,k}^{(t-1)})^m}, \quad 1 \le i \le c$

Step 2   Compute the distance measure $D_{i,k}^2$. The distance to the prototype is calculated based on the fuzzy covariance matrices of the cluster

$$\mathbf{F}_i^{(t)} = \frac{\sum_{k=1}^{N} (\mu_{i,k}^{(t-1)})^m \left(\mathbf{x}_k - \mathbf{v}_i^{(t)}\right) \left(\mathbf{x}_k - \mathbf{v}_i^{(t)}\right)^T}{\sum_{k=1}^{N} (\mu_{i,k}^{(t-1)})^m}, \quad 1 \le i \le c \qquad (A.8)$$

The distance function is chosen as

$$D_{i,k}^2(\mathbf{x}_k, \mathbf{v}_i) = \frac{(2\pi)^{\left(\frac{N}{2}\right)} \sqrt{det(\mathbf{F}_i)}}{\alpha_i} \exp\left(\frac{1}{2} \left(\mathbf{x}_k - \mathbf{v}_i^{(t)}\right)^T \mathbf{F}_i^{-1} \left(\mathbf{x}_k - \mathbf{v}_i^{(t)}\right)\right) \qquad (A.9)$$

with the a priori probability $\alpha_i = \frac{1}{N} \sum_{k=1}^{N} \mu_{i,k}$

Step 3   Update the partition matrix

$$\mu_{i,k}^{(t)} = \frac{1}{\sum_{j=1}^{c} \left(D_{i,k}(\mathbf{x}_k, \mathbf{v}_i) / D_{j,k}(\mathbf{x}_k, \mathbf{v}_j)\right)^{2/(m-1)}}, \quad 1 \le i \le c, 1 \le k \le N \qquad (A.10)$$

Until $||\mathbf{U}^{(t)} - \mathbf{U}^{(t-1)}|| < \varepsilon$.

A.6 Data Sets

A.6.1 Iris Data Set

The Iris data set [17] (http://www.ics.uci.edu) contains measurements on three classes of iris flowers. The data set was made by measurements of sepal length and width and petal length and width for a collection of 150 irises. The analysed data set contains 50 samples from each class of iris flowers (Iris setosa, Iris versicolor and Iris virginica). The problem is to distinguish the three different types of the iris flower. Iris setosa is easily distinguishable from the other two types, but Iris versicolor and Iris virginica are very similar to each other. This data set has been analysed many times to illustrate various clustering methods.

A.6.2 Semeion Data Set

The semeion data set contains 1593 handwritten digits from around 80 persons. Each person wrote on a paper all the digits from 0 to 9, twice. First time in the normal way as accurate as they can and the second time in a fast way. The digits were scanned and stretched in a rectangular box including 16 × 16 cells in a grey scale of 256 values. Then each pixel of each image was scaled into a boolean value using a fixed threshold. As a result the data set contains 1593 sample digits and each digit is characterised with 256 boolean variables. The data set is available form the UCI Machine Learning Repository [18]. The data set in the UCI Machine Learning Repository contains 266 attributes for each sample digit, where the last 10 digits describe the classifications of the digits.

A.6.3 Wine Data Set

The Wine database (http://www.ics.uci.edu) consists of the chemical analysis of 178 wines from three different cultivars in the same Italian region. Each wine is characterised by 13 attributes, and there are 3 classes distinguished.

A.6.4 Wisconsin Breast Cancer Data Set

The Wisconsin breast cancer database (http://www.ics.uci.edu) is a well known diagnostic data set for breast cancer compiled by Dr William H. Wolberg, University of Wisconsin Hospitals [19]. This data set contains 9 attributes and class labels for the

Fig. A.5 Swiss roll data set

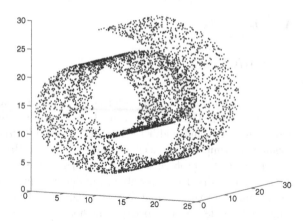

683 instances (16 records with missing values were deleted) of which 444 are benign and 239 are malignant.

A.6.5 Swiss Roll Data Set

The Swiss roll data set is a 3-dimensional data set with a 2-dimensional nonlinearly embedded manifold. The 3-dimensional visualization of the Swiss roll data set is shown in Fig. A.5.

A.6.6 S Curve Data Set

The S curve data set is a 3-dimensional synthetic data set, in which data points are placed on a 3-dimensional 'S' curve. The 3-dimensional visualization of the S curve data set is shown in Fig. A.6.

A.6.7 The Synthetic 'Boxlinecircle' Data Set

The synthetic data set 'boxlinecircle' was made by the authors of the book. The data set contains 7100 sample data placed in a cube, in a refracted line and in a circle. As this data set contains shapes with different dimensions, it is useful to demonstrate the various selected methods. Data points placed in the cube contain random errors (noise), too. In Fig. A.7 data points are yield with blue points and the borders of the points are illustrated with red lines.

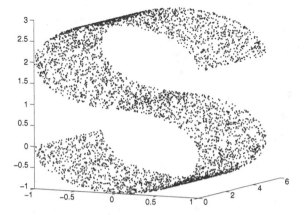

Fig. A.6 S curve data set

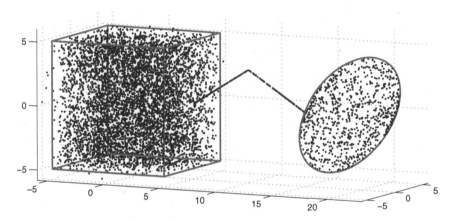

Fig. A.7 'Boxlinecircle' data set

A.6.8 Variety Data Set

The Variety data set is a synthetic data set which contains 100 2-dimensional data objects. 99 objects are partitioned in 3 clusters with different sizes (22, 26 and 51 objects), shapes and densities, and it also contains an outlier. Figure A.8 shows the normalized data set.

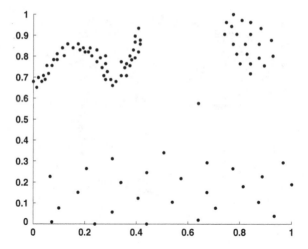

Fig. A.8 Variety data set

Fig. A.9 ChainLink data set

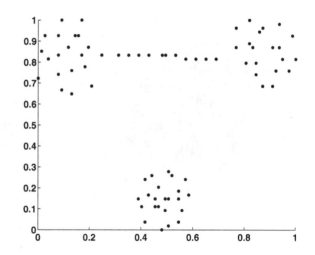

A.6.9 ChainLink Data Set

The ChainLink data set is a synthetic data set which contains 75 2-dimensional data objects. The objects can be partitioned into 3 clusters and a chain link which connects 2 clusters. Hence linkage based methods often suffer from the chaining effect, this example tends to illustrate this problem. Figure A.9 shows the normalised data set.

Fig. A.10 Curves data set

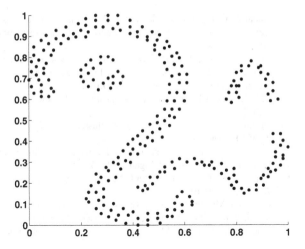

A.6.10 Curves Data Set

The Curves data set is a synthetic data set which contains 267 2-dimensional data objects. The objects can be partitioned into 4 clusters. What makes this data set interesting is that the objects form clusters with arbitrary shapes and sizes, furthermore these clusters lie very near to each other. Figure A.10 shows the normalised data set.

References

1. Jarník, V.: O jistém problému minimálním [About a certain minimal problem]. Práce Moravské Přírodovědecké Společnosti **6**, 57–63 (1930)
2. Prim, R.C.: Shortest connection networks and some generalizations. In. Bell System Technical Journal **36**, 1389–1401 (1957)
3. Kruskal, J.B: On the Shortest Spanning Subtree of a Graph and the Traveling Salesman Problem. In: Proceedings of the American Mathematical Society 7, No. 1, 48–50 (1956).
4. Sneath, P.H.A.: Sokal . Numerical taxonomy. Freeman, R.R. (1973)
5. Nagy, G.: State of the art in pattern recognition. Proceedings of the IEEE **56**(5), 836–862 (1968)
6. King, B.: Step-wise clustering procedures. Journal of the American Statistical Association **69**, 86–101 (1967)
7. Baeza-Yates, R.A.: Introduction to data structures and algorithms related to information retrieval. In Frakes, W.B., Baeza-Yates, R.A. (eds): Information Retrieval: Data Structures and Algorithms, Prentice-Hall, 13–27 (1972).
8. Jain, A., Dubes, R.: Algorithms for Clustering Data. Prentice-Hall (1988).
9. Ward, J.H.: Hierarchical grouping to optimize an objective function. Journal of the American Statistical Association **58**, 236–244 (1963)
10. Zahn, C.T.: Graph-theoretical methods for detecting and describing gestalt clusters. IEEE Transaction on Computers **C20**, 68–86 (1971)
11. Bezdek, J.C., Hathaway, R.J.: VAT: A Tool for Visual Assessment of (Cluster) Tendency. IJCNN **2002**, 2225–2230 (2002)

12. Huband, J., Bezdek, J., Hathaway, R.: Revised Visual Assessment of (Cluster) Tendency (reVAT). Proceedings of the North American Fuzzy Information Processing Society (NAFIPS), 101–104 (2004).
13. Huband, J., Bezdek, J., Hathaway, R.: bigVAT: Visual assessment of cluster tendency for large data sets. Pattern Recognition **38**(11), 1875–1886 (2005)
14. Dempster, A.P., Laird, N.M., Rubin, D.B.: Maximum likelihood from incomplete data via the EM algorithm. Journal of the Royal Statistical Society, Series B (Methodological) **39**, 1–38 (1977)
15. Gustafson, D.E., Kessel, W.C.: Fuzzy clustering with fuzzy covariance matrix. Proceedings of the IEEE CDC, 761–766 (1979).
16. Gath, I., Geva, A.B.: Unsupervised Optimal Fuzzy Clustering. IEEE Transactions on Pattern Analysis and Machine Intelligence **11**, 773–781 (1989)
17. Fisher, R.A.: The Use of Multiple Measurements in Taxonomic Problems. Annals of Eugenics **7**, 179–188 (1936)
18. UC Irvine Machine Learning Repository www.ics.uci.edu/ mlearn/ Cited 15 Oct 2012.
19. Mangasarian, O.L., Wolberg, W.H.: Cancer diagnosis via linear programming. Society for Industrial and Applied Mathematics News **23**(5), 1–18 (1990)

Index

Á. Vathy-Fogarassy and J. Abonyi, *Graph-Based Clustering
and Data Visualization Algorithms*, SpringerBriefs in Computer Science,
DOI: 10.1007/978-1-4471-5158-6, © János Abonyi 2013